Springer Theses

Recognizing Outstanding Ph.D. Research

Aims and Scope

The series "Springer Theses" brings together a selection of the very best Ph.D. theses from around the world and across the physical sciences. Each volume has been selected for its scientific excellence by an internationally renowned institute or department. For greater accessibility to non-specialists, the published versions include an extended introduction, as well as a foreword by the student's supervisor explaining the special relevance of the work for the field. The series as a whole provides a valuable resource both for newcomers to the research fields, and for other scientists seeking detailed background information on special questions. Finally, it serves as an accredited documentation of the valuable contributions made by today's younger generation of scientists.

How to Apply

Theses prepared at internationally leading universities and institutes may be nominated for publication by supervisors with the endorsement of the head of department. Each such department is entitled to nominate one thesis per year. Enquiries and requests for a nomination form should be sent, together with the thesis, to Ms. Nathalie Berg (see link on right).

Eligibility Criteria

To qualify for consideration, theses should fulfil the following criteria:

- They must be written in good English.
- The topic should fall within the confines of Chemistry, Physics, Earth Sciences, Engineering and related interdisciplinary fields such as Materials, Nanoscience, Chemical Engineering, Complex Systems and Biophysics.
- The work reported in the thesis must represent a significant scientific advance.
- A maximum of 30% of the text may be a verbatim reproduction from the author's previous publications.
- They must have been examined and passed during the 12 months prior to nomination.
- The theses should have a clearly defined structure including an introduction accessible to new PhD students and scientists not expert in the relevant field.
- After acceptance, theses should be supplemented by a "Supervisor's Foreword" outlining the wider scientific relevance of the work

Indexed by zbMATH.

Tamanna Jain

Surfing the Gravitational Waves

Analytic Modelling of Gravitational Waves in General Relativity and Beyond

Doctoral Thesis accepted by
University of Cambridge, Cambridge, UK

Springer

Author
Dr. Tamanna Jain
University of Cambridge
Cambridge, UK

Supervisors
Dr. Michalis Agathos
Department of Applied Mathematics
and Theoretical Physics
University of Cambridge
Cambridge, UK

Prof. Ulrich Sperhake
Department of Applied Mathematics
and Theoretical Physics
University of Cambridge
Cambridge, UK

ISSN 2190-5053 ISSN 2190-5061 (electronic)
Springer Theses
ISBN 978-3-031-96179-3 ISBN 978-3-031-96180-9 (eBook)
https://doi.org/10.1007/978-3-031-96180-9

This Springer imprint is published by the registered company Springer Nature Switzerland AG
The registered company address is: Gewerbestrasse 11, 6330 Cham, Switzerland

To my nephew Vanij, that he may know he is dearest to me...

Supervisors' Foreword

Einstein's theory of general relativity, as well as its cousins of modified theories of gravity, predict a phenomenon dubbed *gravitational waves*, disturbances in the fabric of spacetime, which propagate across the universe at the speed of light, very much like wave ripples propagate across a pond of water. Various decade-long international efforts resulted in a network of laser interferometers dedicated to the detection of the tiny displacements of test masses induced by the passage of a gravitational wave. These efforts culminated in the first direct detection of a gravitational-wave signal by LIGO in September 2015, interpreted to have arisen from the coalescence of a black-hole binary about one Giga year away. Over the ensuing decade, about 100 similar events have been observed by the international detector network, composed of LIGO, Virgo and KAGRA.

The dramatic metamorphosis of gravitational-wave physics from a mainly theoretical field of research to an increasingly mature observational science, provides us with a completely new channel to observe the universe, comparable to the invention of X-ray imaging in medicine. Despite the nearly routine accumulation of more and more observational events, the analysis and interpretation of gravitational-wave events remains an exceptionally complex task facing numerous challenges in theory, instrumentation and data analysis. The opportunities and challenges will increase yet further in the future as plans for next-generation detectors—LISA, the Einstein Telescope or Cosmic Explorer—in space and on Earth will unfold. The rewards, on the other hand, will be tremendous, offering insight into the most urgent questions in contemporary physics: from the origin of dark matter and energy to the nature of space and time and the properties of matter under the most extreme conditions.

The interpretation of a gravitational-wave observation bears many similarities with the analysis of a finger print on a crime scene. It relies critically on a catalog of theoretical predictions for the possible physical sources of the event. Just as the comparison of an observed finger print with a data base reveals a suspect, the matching of an observed gravitational-wave signal to the various predictions of waveform catalog informs us about the properties of the astrophysical source. In practice, gravitational-wave observations resemble more closely the measurement of a blurred finger print which, rather than allowing the confident identification of one single suspect, results in a probability distribution across the space of possible culprits. This is due to numerous sources of noise—from ocean waves to shot noise—contaminating the data stream of a gravitational-wave detector. Clearly this kind of analysis, typically employed through the use of Bayesian statistics, requires an accurate and complete data base of gravitational waveform predictions.

The construction of such a data base, a field known as *gravitational-wave source modeling*, employs a vast range of tools including analytic techniques, perturbation theory and numerical methods on high-performance computing systems. The core of this thesis is concerned with the development of more accurate and comprehensive analytic techniques for a particularly common class of gravitational-wave sources, compact binaries involving neutron stars and black holes. Specifically, the thesis pushes analytic calculations inside the so-called *post-Newtonian* and *effective-one-body* framework to higher order of accuracy, both for Einstein's general relativity and scalar-tensor gravity, one of the most popular extensions of general relativity. These developments are complemented by an analysis of one of the most spectacular gravitational-wave events, the neutron-star binary GW170817. By combining this observation with additional information about the neutron-star mass function, tighter constraints on the equation of state of matter at supernuclear densities are derived.

In summary, this thesis pushes the state-of-the-art of analytic gravitational-wave modeling to new levels, further enhancing the waveform catalogs that underlie our interpretation of one of the most intriguing observational programs of our era, *gravitational-wave astronomy*.

Cambridge, UK
December 2024

Dr. Michalis Agathos
Prof. Ulrich Sperhake

Abstract

The first direct detection of gravitational waves (GWs) by LIGO in 2015 opened up new avenues for probing and better understanding the strong-field dynamics of gravity. The continuous improvement in the sensitivity of LIGO, Virgo and now KAGRA along with the forthcoming next-generation detectors like the Einstein Telescope and LISA, will provide a wealth of GW signals to study strong-field dynamics, and to probe the nature of compact objects. Alternative gravitational theories generally predict dynamics different from GR, therefore, it is also expected that the observations of GWs will shed more light on these theories by constraining their parameters. In addition to the applications in usual parameter estimation to improve the constraints, the analytical modelling of waveform generation considered in our work can be used to build search pipelines for the specific beyond GR model considered here.

The detection (searches and parameter estimation) of GWs is based on matching the data observed in the detectors against theoretically predicted waveforms. Hence, highly accurate analytical models are required to construct a bank of waveform templates. The analytical waveform template banks used by LIGO are based on the post-Newtonian (PN) formalism for inspiral and on numerical relativity (NR) for merger and ringdown, or resummation methods such as the effective-one-body (EOB) formalism to describe the complete dynamics. In this thesis, we discuss the construction of analytical waveform models within the EOB formalism for a specific alternative theory of gravity the so-called massless scalar-tensor theory.

More specifically, we describe the map of the dynamics of a binary system in scalar-tensor theory into the EOB formalism by building upon the PN expanded Lagrangian of the two-body problem at 3PN order. We discuss the generalization of the 2PN EOB Hamiltonian to 3PN order for the conservative part of the dynamics and then compute the scalar-tensor corrections to EOB potentials for generic orbits. We also compute the corrections in the EOB Hamiltonian due to tidal effects at 3PN order for circular orbits. Following this, we describe the derivation of the angular momentum flux in the scalar-tensor theory for both the scalar part and the tensorial part up to 1PN and 1.5PN order, respectively. We then use these results along with the energy flux to compute generic orbit radiation reaction effects in the EOB formalism up to 1PN order. Finally, we use the 3PN conservative dynamics results in the EOB formalism to derive the scattering angle in the scalar-tensor theories which can be used in future as a check against the scattering angle computed using scattering amplitude techniques. Finally, we implement up to 3PN conservative dynamics results in the EOB based waveform model TEOBResumS.

Additionally, we investigate the improvement in the inference of the properties of neutron stars using the information of the observation of electromagnetic counterparts of the GW signal GW170817 and improved fits of the threshold mass. The application of the improved fits to the GW170817 signal implies that a significant part of the parameter space is not supported by the observation of the EM counterpart, implying improved constraints on the neutron star radii and hence the equation-of-state.

Acknowledgements First and foremost, I want to thank my supervisor Michalis Agathos for his guidance and support throughout these years and to my mentors and collaborators Donato Bini, Thibault Damour, Anne Davis, Alessandro Nagar, Piero Rettegno and Ulrich Sperhake for many discussions, for encouragement and for their support throughout. I am especially thankful to Thibault Damour for inviting me to visit IHES and to Donato Bini for teaching me to use Maple and cross-verifying my Maple workbooks.

I extend my gratitude to Dhruv Pathak, in whom I found a brother and family outside of my family, for being my biggest cheerleader, and critic, and for more importantly, helping me find ways to enjoy life at all times, thank you for being there!

Lastly, I want to thank my entire family for their endless love and support. Life is easier knowing they are around!

My research at Cambridge has been supported by Cambridge Trust Scholarship.

Competing Interests The author has no competing interests to declare that are relevant to the content of this manuscript.

Conventions and Notations

Units, indices and metric: In this thesis, we will work in the geometric units in which $G = c = 1$, where G is the Newtonian gravitational constant and c is the speed of the light unless specified explicitly.

We use the metric signature $(- + ++)$ to write the action in GR and scalar-tensor theories considered in this thesis. The Greek letters μ, ν, ... as superscripts or subscripts are used to denote the spacetime indices from 0, .., 3 whereas the Latin indices i, j, k, .. as superscripts or subscripts are used to denote spatial indices $1, 2, 3$.

Here, we will also adopt Einstein's summation convention, i.e. the repeated indices are summed over.

4-vector and derivatives: We also follow the usual notations

$$\mathrm{x}^{\mu} = (x^0, \mathbf{x}) = (t, \mathrm{x}),$$

$$\partial_{\mu} = \frac{\partial}{\partial \mathrm{x}^{\mu}},$$

to denote the 4-vectors and their derivatives.

Contents

List of Figures

List of Tables

Chapter 1
Introduction

The theory of General Relativity (GR), developed by Albert Einstein, provides an elegant mathematical framework to explain one of the most common phenomena—gravitation—by revealing that gravity is fundamentally a manifestation of geometry. Essentially, it is a theory of space and time and gravity, where the force of gravity is attributed to the bending and warping of the spacetime. This explains phenomena ranging from falling feathers to the motion of galaxies on the grandest scale and has practical applications in modern technology ranging from relativistic corrections for the accuracy of Global Positioning System (GPS) to telecommunications. The theory has led to various predictions, such as the precession of Mercury, the bending of light, etc., all of which have been experimentally confirmed.

The theory of GR also predicts the existence of gravitational waves (GWs), which are the ripples in the spacetime caused by accelerating masses. These signals originate from highly massive and energetic astrophysical objects such as black holes and neutron stars, where the curvature of the spacetime becomes highly pronounced and hence, are a probe of the strong gravity regime. The first detection of a GW signal caused by the collision of two black holes took place in 2015 by the Laser Interferometry Gravitational-Wave Observatory (LIGO) and VIRGO Collaboration (LVC) [1]. Subsequently, two years later, the GW signal GW170817 from the coalescence of a Binary Neutron Star (BNS) system was observed by the LIGO-Virgo Collaboration [2–4] concurrent with the observation of electromagnetic counterparts [5–7]. The detection of GWs has opened a new window to study the physics of the universe such as cosmic inflation, the formation of the first galaxies, and also to test the theory of GR in extreme gravitational environments with the hope to potentially refine our understanding of gravity.

Although the theory of GR agrees with outstanding precision with the experimental data, there are still regimes where the theory breaks down conceptually, notably on large and small energy scales. At large energy scales, Refs. [8–10] discovered problems of non-renormalizability (i.e., the need for addition of infinite counter terms to cancel divergences) in studying the perturbative expansion of Quantum gravity

T. Jain, *Surfing the Gravitational Waves*, Springer Theses,
https://doi.org/10.1007/978-3-031-96180-9_1

which is required to understand the models of the Big Bang or the vicinity of singularities of black holes. Whereas, at small energy scales, the observation of accelerated expansion of the universe due to some mysterious repulsive force and the presence of more matter than visible matter—dark energy and dark matter—cannot be explained by GR and the standard model of particle physics in its current form [11–16]. These reasons indicate that the theory of gravity needs modification at both small and large energy scales, such that the theory reduces to GR in the intermediate energy scale where the theory has been experimentally verified with remarkable precision.

There are numerous distinct ways in which GR can be modified, making it logical to seek a guiding principle for such modifications. The guiding principle of Lovelock's Uniqueness Theorem [17, 18] found by examining the underlying assumptions of GR states that GR emerges as a unique theory under some assumptions:

In four spacetime dimensions the only divergence-free symmetric rank-2 tensor constructed solely from the metric $g_{\mu\nu}$ *and its derivatives up to second differential order, and preserving diffeomorphism invariance, is the Einstein tensor plus a cosmological term.*

This directly gives us Einstein's Equations. If we assume that the equations of motion follow from a Lagrangian action then the above assumption naturally gives us the Einstein-Hilbert action of GR.

The modifications of GR at these extreme scales can be studied using GWs which may carry the imprints of quantum corrections near the singularity of black holes or the GWs from the early universe may give insight on the conditions right after the Big Bang. This naturally raises the question: *How do we know that a gravitational wave signal has been detected, and how can we determine the properties of its origin from the signal?* The interpretation of a GW signal, like any other observational data, depends on a model built upon the underlying theory governing the system. In the context of gravitational waves, this involves constructing models that predict the appearance of GWs from various astrophysical events, known as waveform templates. Bayesian analysis techniques are then used to infer the properties of these systems based on the waveform templates. This thesis aims to explore the analytical construction of waveform models for a specific modified theory of gravity and to use Bayesian analysis to infer the properties of neutron stars.

The thesis is organised as follows. In this Chapter, we will first briefly review the analytical modelling of compact binaries as a source of GWs within a post-Newtonian (PN) approximation scheme. Then, we will review the effective-one-body (EOB) formalism to generate GW waveform templates for GW detectors, followed by an overview of a specific modified theory called *scalar-tensor gravity* studied in this thesis. As the use of GW detections to explore and understand the dynamics of compact binaries in the strong gravity regime relies on two key components, this thesis is divided into two parts. In Part I, we study the construction of an analytical waveform model within the EOB formalism for scalar-tensor theories where we first derive the EOB dynamics in the conservative sector (i.e. assuming no radiation effects) at 3PN order in Chaps. 2 and 3. Since, the dynamics described within the EOB formalism for the conservative sector, i.e. the EOB Hamiltonian, describe the general binary orbits (quasicircular, elliptic-like, or hyperbolic-like), we use the Hamiltonian derived in

Chaps. 2 and 3 to derive the scattering angle for unbound orbits in scalar-tensor theories in Chap. 4. Following this, we derive the radiated angular momentum flux up to 1.5PN order within the PN approximation for the bound orbit two-body system in Chap. 5 and then derive the corresponding radiation reaction force within the EOB formalism for *generic* orbits up to 1PN order in Chapter 6. In Chap. 7, we used the observation of electromagnetic counterparts of the GW170817 signal to consider the hypothesis that this signal did not promptly collapse to a black hole, and use the mass-ratio dependent fits for the threshold mass to study possible improvements in the inference of NS properties.

1.1 Analytical Modelling of Gravitational Waves

With the recent detections and the hope of more interesting detections, the physics of GWs is in a very exciting period. The GW signals are identified using *matched filtering* techniques to compare the data observed in the detectors against the theoretically predicted waveforms. This naturally leads to the question: *What does the waveform look like, and how do we theoretically predict it?*

A binary system of compact objects goes through three phases during its coalescence; (i) the inspiral phase, when two objects spiral inward and the characteristic orbital velocity is much smaller than the speed of light and the structure of the waveform for non-spinning quasi-circular binaries is such that it has a sinusoidal pattern[1] with increasing amplitude and frequency as the two objects spiral in, (ii) the merger phase, when a collapse of two objects forms a remnant compact object and the waveform reaches its peak amplitude reflecting the violent process during the merger, and (iii) the ringdown phase, when the highly perturbed remnant rings down to a stable BH configuration and the waveform looks like a decaying sinusoid.

The theory of GR is described by the Einstein-Hilbert action,

$$S_{\mathrm{EH}} = \frac{c^4}{16\pi G}\int d^4x\sqrt{-g}R,$$

where c is the speed of light, G is the Newtonian constant, R is the Ricci scalar and g is the determinant of the metric describing the spacetime. This is a complicated, nonlinear, tensorial theory of the gravitational field which makes it difficult to construct physically useful exact solutions. The two particularly famous exact solutions in GR known as Schwarzschild black hole (BH) solution [19] and Kerr BH solution [20] are endowed with high degrees of symmetry and hence their regime of validity is limited. However, the solutions have been proven to be of enormous importance as they describe the spacetime of isolated BHs which are accepted to be found throughout the universe with a notable application that the part of the

[1] The introduction of additional parameters like eccentricity, spins would modify the morphology of the GW signal by modulating the sinsuoidal pattern of amplitude and frequency.

Schwarzschild solution describes the exterior geometry of any static spherical star or planet. But gravitational radiation involves spacetimes that are highly dynamical and nonsymmetrical, and no exact solution of Einstein's equations is known for such a physical configuration which describes the radiation and propagation of GWs from the source. Hence, gravitational radiation is studied analytically using *approximation* schemes such as (i) post-Newtonian approximation where it is assumed that the internal motion of source is slow and the gravitational field of the source is weak and the perturbative expansion is around the Newtonian gravity in powers of v/c, where v is the velocity of source, (ii) post-Minkowskian approximation where it is assumed that the gravitational field of the source is weak and the perturbative expansion is around the Minkowski (flat) spacetime in powers of gravitational constant G and (iii) black-hole perturbation theory where it is assumed that one of the components of the binary system is more massive than the other and the perturbative expansion is around the Kerr metric in powers of perturbing mass (mass-ratio) to solve Einstein's equations. The domain of validity of the post-Newtonian approximation is the spacetime region $0 < r < \mathscr{R}$, where $\mathscr{R}$ is the radial distance defining the *near-zone* and of the post-Minkowskian approximation is $d < r < \infty$, where d is the size of the source. The near-zone is defined as the region where the radial distance r is the less than the typical reduced wavelength of the radiation emitted. The radiated GW in the *far-zone* is then computed by matching the post-Newtonian expansion to the post-Minkowskian expansion in the intermediate regime ($d < r < \mathscr{R}$) where both the approximation schemes are valid [21].

Yet, an alternative method for computationally solving the complete Einstein's equations, known as numerical relativity (NR), has seen substantial advancement in recent years after the breakthrough of the first NR gravitational radiation emitted from the inspiral of compact objects [22–24], which in turn have enhanced our understanding of the features of the gravitational radiation emitted from colliding compact objects. However, due to high computational costs, the NR simulations cover only a patch of the parameter space, implying that the analytical models are required to generate a bank of waveform templates for GW detections. It should also be noted that due to high computational cost NR waveforms are quite short and hence, alone are not sufficient for parameter estimation of sources of GWs using Bayesian inference techniques, implying the need of theoretical models for GW analysis. The theoretical models of the waveform can be matched against the waveform generated using the NR simulations for accuracy and the construction of NR-informed hybrid waveform models to be used for detection and inference of source parameters of GWs. Moreover, the dimensionality of the parameter space is increased by considering modified theories of gravity which introduce theory dependent parameters in the dynamics to encompass the modifications to GR. This becomes a computational bottleneck for NR simulations, and therefore the analytical computations for waveform generation, as done in this thesis, is important.

The simplest approximation to study relativistic corrections due to GR is to consider the linearised theory approximation, i.e., the gravitational fields are extremely weak such that the expansion around a flat spacetime is justified. In this approximation, Einstein's equations have wave-like solutions that can be written in powers of

$\frac{\mathrm{v}}{c}$ where v is the source velocity and is assumed to be independent of the source's field strength. However, in astrophysical systems such as compact object binaries, the velocity and curvature are interdependent, being linked by gravitational forces. This is because, for self-gravitating systems such as compact objects, the virial theorem[2] implies that $\frac{R_S}{d} \approx (\frac{\mathrm{v}}{c})^2$, where R_S is the Schwarzschild radius associated with the total mass of the system and r is the size of the system. Therefore, we should also account for corrections due to deviations from the background spacetime which are measured by $\frac{R_S}{d}$ along with the corrections due to $\frac{\mathrm{v}}{c}$ for gravitational wave signals from these astrophysical sources. Hence, to obtain a more accurate description of gravitational radiation from such sources, the post-Newtonian approximation scheme must be used.

The post-Newtonian (PN) approximation is based on the assumption that the gravitational fields in and around the source are suitably weak (the fields of the propagating waves weaken progressively as they leave the source) and that the motion within the source is slow compared to the speed of light. This implies that a parameter ε can be used to characterise the system as

$$\varepsilon \approx \left(\frac{\mathrm{v}}{c}\right)^2 \approx \left(\frac{GM}{dc^2}\right).$$

The power of $(\mathrm{v}/c)^2$ is used to label the PN order such that from the above definition ε is 1PN order, ε^2 is 2PN order, and so on. This approximation is then used to expand the energy-momentum tensor $T_{\mu\nu}$ due to matter fields and the spacetime metric $g_{\mu\nu}$ in powers of ε entering in Einstein's equations,

$$G_{\mu\nu} = \frac{8\pi G}{c^4} T_{\mu\nu},$$

where $G_{\mu\nu}$ is the Einstein tensor describing the geometry of the spacetime with the metric $g_{\mu\nu}$. The equations thus obtained are then solved order-by-order in ε to obtain the corrections to the metric at higher PN orders (see, Refs. [25–28] for

---

[2] Virial Theorem states that for self-gravitating, stable systems the total kinetic energy is equal to minus half the potential energy

$$\mathrm{K.E} = -\frac{1}{2}U,$$
$$\frac{1}{2}\mu \mathrm{v}^2 = \frac{1}{2}\frac{G\mu M}{r},$$

where $M = m_1 + m_2$ is the total mass of the binary with masses m_1 and m_2, $\mu = \frac{m_1 m_2}{M}$ is the reduced mass of the system and d is the size of system. From the Schwarzschild BH solution, we have $R_S = \frac{2GM}{c^2}$, implying

$$\frac{R_S}{d} \approx \left(\frac{\mathrm{v}}{c}\right)^2,$$

for systems held together by gravitational forces.

detailed reviews of the PN approximation scheme). However, before utilising the PN approximation to generate GW templates, the accuracy of the waveform model must be assessed by comparing the results from the PN expansion with the analytically known exact solution. The GW waveform structure depends sensitively on the frequency at which the binary system transitions from the inspiral phase to the plunge phase followed by the merger. Therefore, the frequency of the innermost stable circular orbit (ISCO), the last stable circular orbit marking the transition from the inspiral phase to the plunge, computed using the PN expansion must be compared with the frequency computed using the exact solution for a test particle (i.e., the mass of particle m_2 is very small such that $m_1 \gg m_2$) moving in the Schwarzschild spacetime.

From the equation of motion of a test particle in the Schwarzschild spacetime, the exact value of the energy for a test particle on a circular orbit in terms of the dimensionless frequency variable $x = (\frac{GM\Omega}{c^3})^{2/3}$ is

$$E(x)_{\text{exact}} = \mu c^2 \frac{1-2x}{\sqrt{1-3x}} - 1. \tag{1.1}$$

Here, $M = m_1 + m_2$ is the total mass, $\mu = \frac{m_1 m_2}{M}$ is reduced mass, and Ω is the orbital frequency.

The energy in the PN formalism up to 3PN order for a comparable mass binary system is [29–36]

$$E(x) = -\frac{\mu c^2 x}{2}\left\{1 + \left(\frac{-3}{4} - \frac{1}{12}\nu\right)x + \left(\frac{-27}{8} + \frac{19}{8}\nu - \frac{1}{24}\nu^2\right)x^2 + \left[\frac{-675}{64} + \left(\frac{209323}{4032} - \frac{205}{96}\pi^2 - \frac{110}{9}\lambda\right)\nu - \frac{155}{96}\nu^2 - \frac{35}{5184}\nu^3\right]x^3 + O(x^4)\right\}, \tag{1.2}$$

where $\nu = \frac{m_1 m_2}{M^2}$ is the symmetric mass ratio and λ is a regularization parameter entering at 3PN order [31, 34, 35].

The location of the ISCO corresponds to the minimum of the energy function $E(x)$, i.e. when $(dE(x)/dx)_{ISCO} = 0$. Now, using Eq. (1.2), we compute the minimum of the Energy and the corresponding orbital frequency for the test particle case $\nu = 0$ and compare this result with the minimum obtained from the exact solution using Eq. (1.1). The comparison of the corresponding ISCO orbital frequencies $\Omega'^n \equiv \frac{\Omega_{\text{nPN}}}{\Omega_{\text{exact}}} = \frac{(x)^{3/2}_{\text{nPN}}}{x^{3/2}_{\text{exact}}}$ is

$$\Omega'^1 = 8.0, \qquad \Omega'^2 = 1.8, \qquad \Omega'^3 = 1.3.$$

Here nPN corresponds to nth PN order.

This comparison suggests that the PN series does not converge fast enough to determine the location of the ISCO which motivated the studies of resummation[3] of the PN results [37–39]. Since, the location of the ISCO is important to understand the global structure of GWs, a new approach, known as the effective-one-body (EOB) formalism, has been developed in Refs. [40, 41] to resum the PN results to extend their validity beyond the weak-field and slow-motion regime. The main aim of Part I of this thesis will be to derive analytical results for a specific modified theory within the EOB formalism. Before proceeding further, we will now review the main result within the EOB formalism for GR which is then extended to modified gravity in Part I.

1.1.1 Effective-One-Body Formalism

The coalescence of a binary system of compact objects is studied in the EOB formalism by mapping the dynamics of the real two-body system into the dynamics of a test mass moving in an effective external spacetime. This analytically describes both the motion and the radiation of coalescing binaries from inspiral through merger to ringdown in contrast to the PN approximation scheme which could only be reliably used for the inspiral phase. The three main features of the EOB approach are: the conservative (Hamiltonian) part of the dynamics (i.e. when radiation effects are not considered), the expression for the radiation-reaction effects, and the description of GWs emitted by the coalescence of a compact binary. Here, we will briefly discuss all three key components of the EOB formalism within GR to generate GWs for detections.

1.1.1.1 Conservative Part of the Dynamics

The basic idea of the EOB formalism is to relate the two-body dynamics to the effective dynamics described by an effective action

$$S_{\rm eff}[\mathrm{x}_{\rm eff}] = -\int \mathrm{m}_{\rm eff}\, \mathrm{c}\, \mathrm{ds}_{\rm eff}, \tag{1.3}$$

for a body of mass $m_{\rm eff}$ moving in a spherically symmetric external spacetime with $\mathrm{x}_{\rm eff}$ as the coordinates of the effective particle (worldline of an effective particle). The line element $ds_{\rm eff}$ for a spherically symmetric spacetime in coordinate system $(r_{\rm eff}, \theta_{\rm eff}, \phi_{\rm eff})$ is

[3] The basic idea of the resummation techniques is to complete the information of a series expansion known for a few terms by using the available information of its global behaviour. This approach aims to improve the accuracy and convergence of the series.

$$ds^2_{\rm eff} = g^{\rm eff}_{\mu\nu} dx^\mu_{\rm eff} dx^\nu_{\rm eff} = -A(r_{\rm eff})c^2 dt^2_{\rm eff} + B(r_{\rm eff})dr^2_{\rm eff} + r^2_{\rm eff}(d\theta^2_{\rm eff} + \sin^2\theta_{\rm eff}\, d\phi^2_{\rm eff}), \tag{1.4}$$

where $A(r_{\rm eff})$ and $B(r_{\rm eff})$ are the two EOB metric potentials, $t_{\rm eff}$ is the coordinate time of the effective EOB metric and $r_{\rm eff}$ is the radial separation in EOB coordinates. In the rest of this section, we specialise in equatorial motion, i.e. set $\theta_{\rm eff} = \pi/2$. The metric potentials $A(r_{\rm eff})$ and $B(r_{\rm eff})$ are constructed as an expansion in $1/(c^2 r_{\rm eff})$

$$A(r_{\rm eff}) = 1 + a_1 \frac{1}{c^2 r_{\rm eff}} + a_2 \left(\frac{1}{c^2 r_{\rm eff}}\right)^2 + a_3 \left(\frac{1}{c^2 r_{\rm eff}}\right)^3 + \cdots,$$
$$B(r_{\rm eff}) = 1 + b_1 \frac{1}{c^2 r_{\rm eff}} + b_2 \left(\frac{1}{c^2 r_{\rm eff}}\right)^2 + b_3 \left(\frac{1}{c^2 r_{\rm eff}}\right)^3 + \cdots. \tag{1.5}$$

As we can see, two mass scales enter into the effective dynamics; (i) the effective mass $m_{\rm eff}$ and (ii) a mass scale $M_{\rm eff}$ in the coefficients a_n's and b_n's as the expansion of the metric potentials $A(r_{\rm eff})$ and $B(r_{\rm eff})$ is dimensionless. The mass scale $M_{\rm eff}$ is defined as

$$a_1 = -2GM_{\rm eff}. \tag{1.6}$$

The EOB action given in Eq. (1.3) is invariant under time translations and space rotations, therefore, it can be separated as,

$$S_{\rm eff} = -\,\mathcal{E}_{\rm eff} t_{\rm eff} + J_{\rm eff}\phi_{\rm eff} + S^r_{\rm eff}(r_{\rm eff}), \tag{1.7}$$

where $\mathcal{E}_{\rm eff}$ and $J_{\rm eff}$ are the energy and the angular momentum flux in the EOB formalism and $S^r_{\rm eff}(r_{\rm eff})$ is the effective radial action. By requiring that the dynamics of the effective particle in the effective metric is described by the mass-shell condition $p_\mu p^\mu + m^2_{\rm eff}c^2 = 0$ and then solving the corresponding Hamilton-Jacobi equation,

$$g^{\mu\nu}_{\rm eff} \frac{dS_{\rm eff}}{dx^\mu_{\rm eff}} \frac{dS_{\rm eff}}{dx^\nu_{\rm eff}} + m^2_{\rm eff}c^2 = 0, \tag{1.8}$$

using the separated action, we obtain,

$$-\frac{1}{A(r_{\rm eff})}\frac{\mathcal{E}^2_{\rm eff}}{c^2} + \frac{1}{B(r_{\rm eff})}\left(\frac{dS^r_{\rm eff}(r_{\rm eff})}{dr_{\rm eff}}\right)^2 + \frac{J^2_{\rm eff}}{r^2_{\rm eff}} + m^2_{\rm eff}c^2 = 0. \tag{1.9}$$

The solution of the above equation computes the radial action variable $I^{r_{\rm eff}}_{\rm eff} = \frac{2}{2\pi}\int dr_{\rm eff}\frac{dS^r_{\rm eff}}{dr_{\rm eff}}$ as a function of $\mathcal{E}_{\rm eff}$ and $J_{\rm eff}$, which is then inverted to obtain $\mathcal{E}_{\rm eff}$ as a function of effective action variables $N_{\rm eff}$ and $J_{\rm eff}$, where $N_{\rm eff} = I^{r_{\rm eff}}_{\rm eff} + J_{\rm eff}$. The explicit expression of $\mathcal{E}_{\rm eff}$ up to 2PN in GR is given in Ref. [40].

Now, to map the EOB Hamiltonian derived following the above method to the Hamiltonian of the two-body system, Ref. [40] proposed to consider the problem in terms of quantum mechanics which naturally gives a one-to-one correspondence between the action variables of the effective Hamiltonian ($N_{\text{eff}}, J_{\text{eff}}$) and the two-body Hamiltonian ($N_{\text{real}}, J_{\text{real}}$), i.e. $N_{\text{eff}} = N_{\text{real}}$ and $J_{\text{eff}} = J_{\text{real}}$. However, there is no one-to-one correspondence between the energy levels of the real two-body dynamics and EOB as the rest mass contributions are different for both energy levels.[4] Reference [40] proposed a one-to-one map between the two-body and the EOB energy as $\mathscr{E}_{\text{eff}} = f[\mathscr{E}_{\text{real}}]$ such that

$$\frac{\mathscr{E}_{\text{eff}}}{m_{\text{eff}}c^2} - 1 = f\left(\frac{E_{\text{real}}^{\text{nonrel}}}{\mu c^2}\right) = \frac{E_{\text{real}}^{\text{nonrel}}}{\mu c^2}\left(1 + \alpha_1 \frac{E_{\text{real}}^{\text{nonrel}}}{\mu c^2} + \alpha_2 \left(\frac{E_{\text{real}}^{\text{nonrel}}}{\mu c^2}\right)^2 + \cdots\right), \tag{1.10}$$

where $E_{\text{real}}^{\text{nonrel}}$ is the non-relativistic energy of the two-body system and α_n's are the unknown coefficients to be determined from the energy map.

The unknown coefficients entering the energy map are uniquely determined by imposing two conditions. The first condition is to assume that the mass of the effective particle m_{eff} coincides with the reduced mass μ. The second is to assume that the linear term (Gm_1m_2) ("one-graviton" exchange) in the effective metric corresponds to the Schwarzschild metric (of mass M), i.e. $a_1 = -2GM$ and $b_1 = 2GM$. The second is a natural condition as the effective potential due to one-graviton-exchange depends only on the relative velocity, which in the test mass limit $\nu \to 0$ coincides with the effective potential of the linearised (i.e. assuming weak-field) Schwarzschild metric. Hence, the effective metric in the EOB formalism is a ν-deformation of the Schwarzschild metric and reproduces at all orders in G the interactions generated by the exchange of gravitons.

Inserting the two-body system and EOB Hamiltonian from Ref. [40] in Eq. (1.10) yields a system of equations to obtain the unknown coefficients a_n, b_n, and α_n. The solutions of these coefficients up to 2PN order presented in GR are

$$a_2 = 0, \quad a_3 = 2\nu(GM)^3, \quad b_2 = (4 - 6\nu)(GM)^2, \quad \alpha_1 = \frac{\nu}{2}, \quad \alpha_2 = 0. \tag{1.11}$$

Inserting the coefficients in the relation (1.10), the mapping from the two-body energy $E_{\text{real}}^{\text{nonrel}}$ to the EOB energy $\mathscr{E}_{\text{eff}}$ is given by

$$\frac{\mathscr{E}_{\text{eff}}}{\mu c^2} = \frac{(E_{\text{real}}^{\text{nonrel}})^2 - m_1^2c^4 - m_2^2c^4}{2m_1m_2c^4}. \tag{1.12}$$

The dynamics of a test particle in the effective metric is studied noting that the Hamilton-Jacobi equation given in Eq. (1.9) yields the relativistic effective Hamiltonian

$$H_{\text{eff}} = \mu c^2 \sqrt{A(r_{\text{eff}})\left[1 + \frac{p_{\text{r,eff}}^2}{\mu^2 c^2 B(r_{\text{eff}})} + \frac{J_{\text{eff}}^2}{\mu^2 c^2 r_{\text{eff}}^2}\right]}, \tag{1.13}$$

[4] The rest mass contribution for the real two-body system is Mc^2 whereas in the EOB formalism it is $m_{\text{eff}}c^2$.

where $p_{r,\mathrm{eff}} = \frac{\partial S_{r,\mathrm{eff}}}{\partial r_{\mathrm{eff}}}$ is the radial momentum. This parallels exactly the analysis done for the Schwarzschild metric. Setting the radial momentum $p_{r,\mathrm{eff}} = 0$ for circular orbits, the position, frequency, angular momentum and energy of the ISCO in the effective dynamics is studied by deriving the minimum of Eq. (1.13). The effective quantities derived are only mathematical quantities to simplify the analysis by combining all the PN information into two metric potentials $(A(r_{\mathrm{eff}}), B(r_{\mathrm{eff}}))$ and are not the physical quantities of the two-body system. For physical purposes, the effective quantities are transformed into real physical quantities of the two-body system. For energy, this is achieved by inverting Eq. (1.12) as

$$E^{\mathrm{rel}}_{\mathrm{real}} = Mc^2 \sqrt{1 + 2\nu \left(\frac{\mathscr{E}_{\mathrm{eff}} - \mu c^2}{\mu c^2} \right)}, \tag{1.14}$$

where $E^{\mathrm{rel}}_{\mathrm{real}} = E^{\mathrm{nonrel}}_{\mathrm{real}} + Mc^2$. This relation between the two-body and EOB Hamiltonian was then proved to be correct up to 3PN order in GR [42] by explicitly computing the unknown coefficient α_n entering at 3PN order. In Ref. [42], it has also been shown that for simplifying the mapping between the real two-body and EOB Hamiltonian at higher PN orders (i.e. at 3PN order and above), the Hamilton-Jacobi equation for the effective particle given in Eq. (1.8) needs to be generalised by adding *non-geodesic* terms such that the dynamics of the effective particle is described by the "effective" Hamilton-Jacobi equation of the form

$$g^{\mu\nu}_{\mathrm{eff}} \frac{dS_{\mathrm{eff}}}{dx^{\mu}_{\mathrm{eff}}} \frac{dS_{\mathrm{eff}}}{dx^{\nu}_{\mathrm{eff}}} + m^2_{\mathrm{eff}} c^2 + Q\left(\frac{dS_{\mathrm{eff}}}{dx_{\mathrm{eff}}} \right) = 0, \tag{1.15}$$

where the non-geodesic term Q is at least quartic in momentum (p^{eff}_{μ}) of the effective particle. Hence, the 2PN order effective Hamiltonian given in Eq. (1.13) is also generalised by the addition of non-geodesic terms starting at 3PN order which we discuss in Chapter 2.

Recently this relation was proven to be true for *all* PN orders within the post-Minkowskian scheme in Ref. [43] for both GR and a class of modified theories of gravity called scalar-tensor theories.

1.1.1.2 Radiation Reaction Effects and Waveforms

In the previous subsection, we have discussed the incorporation of the two-body dynamics into the dynamics of an "effective" particle for the conservative part within the EOB formalism. Here, we will briefly discuss the resummed expressions that describe radiation reaction effects [41] and the waveform emitted at infinity [44]. Most of the compact binary systems are expected to lose their initial eccentricity and circularise due to gravitational damping, therefore, as a first step, we study the radiation reaction effects for quasi-circular orbits within the EOB approach [41, 45]. With the recent advancements in GW detectors and the possibility of future detectors, there is now hope that eccentricity might be observed soon. The radiation reaction effects for the *general* orbits will be discussed in Chapter 6.

Here, we consider the equations of motion in the Hamiltonian formalism. More precisely, we assume the motion is in the equatorial plane, i.e., set $\theta_{\rm eff} = \frac{\pi}{2}$) and work in a polar coordinate system with phase space variables $r_{\rm eff}, \phi_{\rm eff}, p_{\rm r,eff}, p_{\phi,\rm eff}$ where $p_{\rm r,eff}$ is the radial momentum and $p_{\phi,\rm eff} = J_{\rm eff}$ is the angular momentum. For quasi-circularised binaries, it is enough to consider the radiation reaction force in the equation of motion of $p_{\phi,\rm eff}$ only, as for the quasi-circularised orbits $p_{\rm r,eff} \ll p_{\phi,\rm eff}$ and hence, the radiation reaction force in the equation of motion of $p_{\rm r,eff}$ can be neglected. The equations of motion corresponding to the two-body Hamiltonian $H_{\rm real}(r_{\rm eff}, \phi_{\rm eff}, p_{\rm r,eff})$ expressed in terms of the effective variables after including the radiation reaction force[5] in the Hamiltonian formalism are,

$$\frac{d\phi_{\rm eff}}{dt_{\rm eff}} = \frac{\partial H^{\rm real}}{\partial p_{\phi,\rm eff}} = \Omega, \tag{1.16}$$

$$\frac{dr_{\rm eff}}{dt_{\rm eff}} = \frac{\partial H^{\rm real}}{\partial p_{\rm r,eff}}, \tag{1.17}$$

$$\frac{dp_{\phi,\rm eff}}{dt_{\rm eff}} = \mathscr{F}_{\phi}, \tag{1.18}$$

$$\frac{dp_{\rm r,eff}}{dt_{\rm eff}} = -\frac{\partial H^{\rm real}}{\partial r_{\rm eff}}, \tag{1.19}$$

where $\mathscr{F}_{\phi}$ is the damping force due to the radiation effects. Then using the map between the two-body and the EOB Hamiltonian,

$$\hat{H}^{\rm real} = \frac{H^{\rm real}}{\mu} = \frac{1}{\nu}\sqrt{1 - 2\nu(\hat{H}_{\rm eff} - 1)}, \tag{1.20}$$

we obtain,

$$\frac{d\phi_{\rm eff}}{d\hat{t}_{\rm eff}} = \frac{\hat{p}_{\phi,\rm eff}A(\hat{r}_{\rm eff})}{\nu\hat{r}_{\rm eff}^2\hat{H}^{\rm real}\hat{H}_{\rm eff}}, \tag{1.21}$$

$$\frac{d\hat{r}_{\rm eff}}{d\hat{t}_{\rm eff}} = \frac{\hat{p}_{\rm r,eff}A(\hat{r}_{\rm eff})}{\nu B(\hat{r}_{\rm eff})\hat{H}^{\rm real}\hat{H}_{\rm eff}}, \tag{1.22}$$

---

[5] The equation of motion corresponding to the action (1.3) in Lagrangian dynamics is

$$\frac{d}{dt}\frac{\partial L}{\partial \dot{x}_i} - \frac{\partial L}{\partial x_i} = 0,$$

with $x_i = r_{\rm eff}, \phi_{\rm eff}$. Then by adding the radiation reaction force (damping force $\mathscr{F}_i$) the equation of motion becomes

$$\frac{d}{dt}\frac{\partial L}{\partial \dot{x}_i} - \frac{\partial L}{\partial x_i} = \mathscr{F}_i.$$

$$\frac{d\hat{p}_{\phi,\mathrm{eff}}}{d\hat{t}_{\mathrm{eff}}} = \hat{\mathscr{F}}_\phi, \tag{1.23}$$

$$\frac{d\hat{p}_{\mathrm{r,eff}}}{d\hat{t}_{\mathrm{eff}}} = -\frac{1}{2\nu\hat{H}^{\mathrm{real}}\hat{H}_{\mathrm{eff}}}\left[A' + \frac{\hat{p}^2_{\phi,\mathrm{eff}}}{\hat{r}^2_{\mathrm{eff}}}\left(A' - \frac{2A}{\hat{r}_{\mathrm{eff}}}\right) + \hat{p}^2_{\mathrm{r,eff}}\left(\frac{A'}{B} - \frac{AB'}{B^2}\right)\right], \tag{1.24}$$

where $A' = \frac{dA}{dr_{\mathrm{eff}}}$ and $B' = \frac{dB}{dr_{\mathrm{eff}}}$. Here, we use *hat* to denote dimensionless coordinates as

$$\hat{t}_{\mathrm{eff}} = \frac{t_{\mathrm{eff}}}{GM}, \quad \hat{r}_{\mathrm{eff}} = \frac{r_{\mathrm{eff}}}{GM}, \quad \hat{p}_{\mathrm{r,eff}} = \frac{p_{\mathrm{r,eff}}}{\mu}, \quad \hat{p}_{\phi,\mathrm{eff}} = \frac{p_{\phi,\mathrm{eff}}}{\mu GM}, \quad \hat{H}_{\mathrm{eff}} = \frac{H_{\mathrm{eff}}}{\mu}. \tag{1.25}$$

The radiation reaction force $\hat{\mathscr{F}}_\phi$ must be replaced by a resummed factored flux function for continuity with the test mass case $\nu \to 0$ within the EOB formalism. The various methods to resum the radiation reaction force $\hat{\mathscr{F}}_\phi$ are proposed in Refs. [44, 46, 47] (see, Ref. [45] for the overview). The generalisation of the quasi-circular approximation is studied in Refs. [48, 49] for GR and Chapter 6 for a specific modified theory of gravity.

The gravitational wave signal emitted in the inspiral and plunge phase in the simplest approximation can be computed by solving numerically the equations of motion and then applying the quadrupolar formula to obtain the GW strain [50]. References [44, 46, 47] presented a multiplicative approach to decompose the relativistic multipolar waveform $h_{\ell m}$ with modes (ℓ, m) as a product of various contributions, that is,

$$h^{(\tilde{\varepsilon})}_{\ell m} = h^{(N,\tilde{\varepsilon})}_{\ell m}\hat{h}^{(\tilde{\varepsilon})}_{\ell m}. \tag{1.26}$$

Here, the first term $h^{(N,\tilde{\varepsilon})}_{\ell,m}$ is the leading order Newtonian contribution to waveform and the second term $\hat{h}^{(\tilde{\varepsilon})}_{\ell m}$ is the product of all the remaining contributions. As all the remaining contributions represent the resummed version of the PN corrections, the structure of $\hat{h}^{(\tilde{\varepsilon})}_{\ell m}$ should be of type $1 + \mathscr{O}(x)$.

The Newtonian contribution to circular orbits is

$$h^{(N,\tilde{\varepsilon})}_{\ell m} = \frac{GM\nu}{c^2 r_{\mathrm{eff}}} n^{(\tilde{\varepsilon})}_{\ell m} c_{\ell+\tilde{\varepsilon}}(\nu)\, x^{\ell+\tilde{\varepsilon}}\, Y^{\ell-\tilde{\varepsilon},-m}, \tag{1.27}$$

where, $\tilde{\varepsilon}$ denotes the parity of $\ell + m$ such that $\tilde{\varepsilon} = 0$ for "mass-generated" (even) multipoles and $\tilde{\varepsilon} = 1$ for "current-generated" (odd) multipoles, $n^{(\tilde{\varepsilon})}_{\ell m}$ and $c_{\ell+\tilde{\varepsilon}}$ are numerical coefficients and $Y^{\ell,m}(\theta, \phi)$ are scalar spherical harmonics. The remaining PN correction factor $\hat{h}^{(\varepsilon)}_{\ell m}$ is further decomposed into the product of four factors,

$$\hat{h}^{(\tilde{\varepsilon})}_{\ell m} = \hat{S}^{\tilde{\varepsilon}}_{\mathrm{eff}}\, T_{\ell m}\, e^{\mathrm{i}\delta_{\ell m}}\, \rho^{\ell}_{\ell m}, \tag{1.28}$$

where $\hat{S}_{\rm eff}^{(\tilde{\varepsilon})}$ is a μ normalised effective source, $T_{\ell m}$ is a resummation [46, 47] of an infinite number of "leading logarithms" entering the tail effects [51, 52], $\delta_{\ell m}$ is a supplementary phase to incorporate the phase effects not included in the complex tail factor of $T_{\ell m}$, and $\rho_{\ell m}^{\ell}$ is the ℓth-power of a building block $\rho_{\ell m}$ introduced in Ref. [44]. After combining the above equations in Eq. (1.26), the multipolar waveform for a circular nonspinning compact binary is

$$h_{\ell m}^{(\tilde{\varepsilon})} = \frac{GM\nu}{c^2 r} n_{\ell m}^{(\tilde{\varepsilon})} c_{\ell+\tilde{\varepsilon}}(\nu)\, x^{\ell+\tilde{\varepsilon}}\, Y^{\ell-\tilde{\varepsilon},-m}\, \hat{S}_{\rm eff}^{\tilde{\varepsilon}}\, T_{\ell m}\, e^{{\rm i}\delta_{\ell m}}\, \rho_{\ell m}^{\ell}. \tag{1.29}$$

This newly resummed waveform is then used to define the resummation of the *radiation reaction* force $\mathscr{F}_\phi$ as

$$\mathscr{F}_\phi = \frac{1}{\Omega}\left[\frac{2}{16\pi G}\sum_{\ell=2}^{\ell=\ell_{\rm max}}\sum_{m=1}^{m=\ell}(M\Omega)^2\left|r_{\rm eff}h_{\ell m}\right|^2\right]. \tag{1.30}$$

The Eqs. (1.29) and (1.30) define the resummed version of the waveform and radiation reaction effects within the EOB formalism. These, together with the information for the conservative part of the dynamics, *complete* the initial value problem within EOB formalism which is then solved numerically to give predictions for the radiation reaction effects and waveform. The waveform model based on the EOB formalism used for the construction of template banks for searches of GWs is `SEOBNR` [53, 54] and EOB based waveform models used in the parameter estimation of GWs are `TEOBResumS` [55–61] and `SEOBNR`.

In the remainder of this thesis, we suppress the subscript eff used to denote effective particle variables for simplicity.

1.2 Scalar-Tensor Theories

The GW detections can be used to test modified theories of gravity on extreme scales where we expect deviations from GR.

Now, to test a modified theory in the strong gravity regime using GWs, we assume that the modifications must be such that the modified theory (i) possesses a well-posed initial value problem, i.e. there exists a unique solution as well as continuous dependence of the solution on the initial conditions, (ii) passes the intermediate tests of GR, and (iii) be simple enough to allow us to make calculable predictions of the structure and dynamics of compact objects and the gravitational radiation they emit so that it can be tested with observational data.

One of the simplest extensions of GR which can be studied using GWs is scalar-tensor gravity (see Ref. [62] for an overview of the extensions of GR). These theories incorporate one or more scalar degrees of freedom into the gravitational sector through nonminimal coupling, where the Ricci scalar in the Einstein-Hilbert action is multiplied by a scalar field(s) dependent function. A scalar field with nonminimal couplings arises in several contexts, such as in string theory [63], in Kaluza-Klein-like theories [64] and also have important applications in cosmology [65]. Hence, the scalar-tensor theory also provides a good framework for studying the phenomenological aspects of various modified theories.

The action for a class of scalar-tensor theories with a single massless scalar field $\phi^{\rm ST}$ coupled to the metric $\tilde{g}_{\mu\nu}$ is

$$S_{\rm ST} = \frac{c^4}{16\pi G}\int d^4x\sqrt{-\tilde{g}}\left(\phi^{\rm ST}\tilde{R} - \frac{\omega(\phi^{\rm ST})}{\phi^{\rm ST}}\tilde{g}^{\mu\nu}\partial_\mu\phi^{\rm ST}\partial_\nu\phi^{\rm ST}\right) + S_{\rm mt}[\Psi_{\rm m}, \tilde{g}_{\mu\nu}], \tag{1.31}$$

where $\tilde{R}$ and $\tilde{g}$ are respectively the Ricci scalar and the determinant of the metric $\tilde{g}_{\mu\nu}$, ω is some function of the scalar field and $S_{\rm mt}$ is the matter part of action with $\Psi_{\rm m}$ collectively denoting matter fields. The function $\omega(\phi^{\rm ST})$ uniquely fixes the scalar-tensor theory and is often called the coupling function. This action is usually referred to as the *Jordan frame* or the *metric frame* action as the matter fields do not directly couple to the scalar field. This implies that the equations of motion for the matter fields do not involve the scalar field explicitly and hence, the test particles follow the geodesics of the metric $\tilde{g}_{\mu\nu}$. Since, the geodesic paths of test particles are dictated by the metric, the trajectories of the freely falling test particles are independent of their mass composition implying that the scalar-tensor theory respects the weak-equivalence principle.[6] Even though the scalar field dependence in the matter action is not explicit these theories do violate the strong equivalence principle because of the scalar field dependence of the mass of a body in these theories. This can be understood as follows.

In Refs. [66, 67], it has been shown that the gravitational constant depends on the value of the scalar field for scalar-tensor theories. The energy of a system bounded by gravitational forces is fundamentally dependent on the gravitational constant, implying that the energy varies with the scalar field. The total energy of the system is related to the inertial mass as $E = mc^2$, where m is the mass of the object, which further implies that the inertial mass of the self-gravitating objects depends on the scalar field and therefore following Ref. [68] the total mass of each body can be described by the mass function $\tilde{m}(\phi^{\rm ST})$ dependent on the value of the scalar field $(\phi^{\rm ST})$. We will use this approach to "skeletonize" the extended objects as point particles suggested in Ref. [68] and justified in Refs. [69, 70] to describe the matter action in scalar-tensor theories (see, Eq. (2.2) of Chap. 2). As a consequence of the scalar field dependence of the mass $\tilde{m}(\phi^{\rm ST})$, the field equations in scalar-tensor theories will depend on the "sensitivity" of the mass of each body to variations in the scalar field, where the sensitivity of a body is defined as

$$s = \frac{d\ln\tilde{m}(\phi^{\rm ST})}{d\ln\phi^{\rm ST}}. \tag{1.32}$$

[6] The weak equivalence principle states that in a uniform gravitational field all freely falling objects follow the same trajectories independent of their composition.

The sensitivity for each body in a binary system can be defined accordingly.

In Ref. [71], it was first pointed out that for theories containing both a metric tensor field and a scalar field, a new metric can be defined such that it differs from the 'physical' metric by a conformal factor depending on the scalar field, i.e., using some suitable conformal transformation of the metric the Einstein-Hilbert term can be separated from the scalar field of the action given in Eq. (1.31). The conformal transformation for scalar-tensor theories as shown in Refs. [69] is

$$\tilde{g}_{\mu\nu} = \mathscr{A}(\phi^{\mathrm{ST}})^2 g_{\mu\nu}, \tag{1.33}$$

where $\mathscr{A}(\phi^{\mathrm{ST}})$ is the conformal factor that incorporates the scalar-tensor theory in the new frame referred to as the *Einstein frame* and $g_{\mu\nu}$ is the Einstein frame metric. The action for the scalar-tensor theory can then be derived using this conformal transformation and a redefinition of the scalar field ϕ^{ST} in terms of the rescaled scalar field φ. The action and field equations for a particular choice of the conformal factor $\mathscr{A}(\phi^{\mathrm{ST}}) = 1/\phi^{\mathrm{ST}}$ and redefined scalar field $\varphi = \phi^{\mathrm{ST}}/\phi_0^{\mathrm{ST}}$, where ϕ_0^{ST} is the value of the scalar field at spatial infinity is given in Refs. [72–77]. Here, we follow the conventions and notations of Ref. [69] to express the action for a single massless scalar field in the Einstein frame,

$$S_{\mathrm{ST}}^{\mathrm{EF}} = \frac{c^4}{16\pi G} \int d^4x \sqrt{-g}\left(R - 2g^{\mu\nu}\partial_\mu\varphi\partial_\nu\varphi\right) + S_{\mathrm{mt}}[\Psi_{\mathrm{m}}, \mathscr{A}(\varphi)^2 g_{\mu\nu}], \tag{1.34}$$

where R is the Ricci scalar in the Einstein frame, φ is the scalar field and Ψ_m collectively denotes the matter fields. The matter action directly couples to the scalar field, therefore, the dynamics of the scalar field φ arise from its coupling to the matter fields Ψ_m. The free dynamics of the Einstein frame metric $g_{\mu\nu}$ is governed by the usual Einstein-Hilbert action. Therefore, in the Einstein frame, the scalar-tensor theory is uniquely fixed when the function $\mathscr{A}(\varphi)$ is defined, and when $\mathscr{A}(\varphi) = \text{const.}$, GR is recovered.

For scalar-tensor theories, the important violations to GR arise due to the non-perturbative phenomena related to the "scalarization" of the neutron stars that take place either in the early or the late stages of the evolution of the binary dynamics known as spontaneous or dynamical scalarization, respectively. The scalarization can be interpreted as a phase transition as the scalar configuration appears only when certain quantities that describe the self gravitating objects such as scalar charge is beyond a threshold (see, Ref. [78] for review on different spontaneous scalarization mechanisms). The spontaneous scalarization for Damour Esposito-Farese (DEF) model of scalar-tensor theories was first studied in Refs. [79, 80]. Currently, the most stringent constraints to the non-perturbative strong-field effects in scalar-tensor theories are given by observations of pulsars by Pulsar Timing arrays [81, 82].

As mentioned previously, theoretical waveform templates are required to detect GWs. This implies that theoretical waveform templates must be constructed for scalar-tensor theories as well. Despite the slow convergence of the PN formalism, it forms an important step in studying the dynamics and waveform generation for compact binaries. Due to its simplicity, there has been an increasing effort to analytically solve the two-body problem in scalar-tensor theories for both the dynamics [73–77, 83] and the waveform generation [84, 85]. However, as argued initially, the results in the PN formalism must be resummed using various techniques given in Refs. [37–39] or mapped to EOB formalism to extend the regime of validity to reliably construct waveform templates. In Part I of the thesis, we aim to map the PN results into the EOB formalism for scalar-tensor theories.

References

1. Abbott BP et al (2016) (Virgo, LIGO Scientific), Observation of gravitational waves from a binary black hole merger. Phys Rev Lett 116:061102. arXiv:1602.03837. https://doi.org/10.1103/PhysRevLett.116.061102 [gr-qc]
2. Abbott BP et al (2017) (LIGO Scientific, Virgo), GW170817: observation of gravitational waves from a binary neutron star inspiral. Phys Rev Lett 119:161101. arXiv:1710.05832. https://doi.org/10.1103/PhysRevLett.119.161101 [gr-qc]
3. Abbott BP et al. (2019) (LIGO Scientific, Virgo), Properties of the binary neutron star merger GW170817. Phys Rev X 9:011001. arXiv:1805.11579. https://doi.org/10.1103/PhysRevX.9.011001 [gr-qc]
4. Abbott BP et al (2019) (LIGO Scientific, Virgo), GWTC-1: a gravitational-wave transient catalog of compact binary mergers observed by LIGO and virgo during the first and second observing runs. Phys Rev X 9:031040. arXiv:1811.12907. https://doi.org/10.1103/PhysRevX.9.031040 [astro-ph.HE]
5. Abbott BP et al (2017) (LIGO Scientific, Virgo, Fermi-GBM, INTEGRAL), Gravitational waves and gamma-rays from a binary neutron star merger: GW170817 and GRB 170817A. Astrophys J Lett 848:L13. arXiv:1710.05834. https://doi.org/10.3847/2041-8213/aa920c [astro-ph.HE]
6. Troja E et al (2017) The X-ray counterpart to the gravitational wave event GW 170817. Nature 551:71–74. arXiv:1710.05433. https://doi.org/10.1038/nature24290 [astro-ph.HE]
7. Abbott BP et al (2017) (LIGO Scientific, Virgo, Fermi GBM, INTEGRAL, IceCube, AstroSat Cadmium Zinc telluride imager team, IPN, Insight-Hxmt, ANTARES, Swift, AGILE Team, 1M2H Team, Dark energy camera GW-EM, DES, DLT40, GRAWITA, Fermi-LAT, ATCA, ASKAP, Las Cumbres observatory Group, OzGrav, DWF (Deeper Wider Faster Program), AST3, CAASTRO, VINROUGE, MASTER, J-GEM, GROWTH, JAGWAR, CaltechNRAO, TTU-NRAO, NuSTAR, Pan-STARRS, MAXI Team, TZAC Consortium, KU, Nordic Optical Telescope, ePESSTO, GROND, Texas Tech University, SALT Group, TOROS, BOOTES, MWA, CALET, IKI-GW Follow-up, H.E.S.S., LOFAR, LWA, HAWC, Pierre Auger, ALMA, Euro VLBI Team, Pi of Sky, Chandra Team at McGill University, DFN, ATLAS Telescopes, High Time Resolution Universe Survey, RIMAS, RATIR, SKA South Africa/MeerKAT) Multi-messenger observations of a binary neutron star merger. Astrophys J Lett 848:L12. arXiv:1710.05833. https://doi.org/10.3847/2041-8213/aa91c9 [astro-ph.HE]
8. Deser S, Van Nieuwenhuizen P, Boulware D (1975) in 7th international conference on gravitation and relativity, pp 1–18
9. Trends in elementary particle theory: international summer institute on theoretical physics in Bonn 1974, Springer Berlin Heidelberg (1975). https://doi.org/10.1007/3-540-07160-1
10. Balian R, Zinn-Justin J (1981) Methods in field theory: Les Houches session XXVIII, Les Houches summer school proceedings series. North-Holland Publishing Company. https://books.google.co.uk/books?id=FIiqzA9YIxEC

11. Zwicky F (1933) Die Rotverschiebung von extragalaktischen Nebeln. Helv Phys Acta 6:110–127. https://doi.org/10.1007/s10714-008-0707-4
12. Weinberg S (1989) The cosmological constant problem. Rev Mod Phys 61:1–23 (1989). https://doi.org/10.1103/RevModPhys.61.1
13. Martin J (2012) Everything you always wanted to know about the cosmological constant problem (But Were Afraid To Ask). Comptes Rendus Physique 13:566–665. arXiv:1205.3365. https://doi.org/10.1016/j.crhy.2012.04.008 [astro-ph.CO]
14. Joyce A, Jain B, Khoury J, Trodden M (2015) Beyond the cosmological standard model. Phys Rep 568:1–98. arXiv:1407.0059. https://doi.org/10.1016/j.physrep.2014.12.002 [astro-ph.CO]
15. Frusciante N, Perenon L (2020) Effective field theory of dark energy: a review. Phys Rep 857:1–63. arXiv:1907.03150. https://doi.org/10.1016/j.physrep.2020.02.004 [astro-ph.CO]
16. Salucci P, Turini N, Di Paolo C (2020) Paradigms and scenarios for the dark matter phenomenon. Universe 6:118. arXiv:2008.04052. https://doi.org/10.1142/9789811233913_0075 [astro-ph.CO]
17. Lovelock D (1971) The Einstein tensor and its generalizations. J Math Phys 12:498–501. https://doi.org/10.1063/1.1665613
18. Lovelock D (1972) The four-dimensionality of space and the Einstein tensor. J Math Phys 13:874–876. https://doi.org/10.1063/1.1666069
19. Schwarzschild K (1916) On the gravitational field of a sphere of incompressible fluid according to Einstein's theory. Sitzungsber Preuss Akad Wiss Berlin (Math Phys) 1916:424–434. arXiv:physics/9912033
20. Kerr RP (1963) Gravitational field of a spinning mass as an example of algebraically special metrics. Phys Rev Lett 11:237–238. https://doi.org/10.1103/PhysRevLett.11.237
21. Blanchet L, Damour T (1992) Hereditary effects in gravitational radiation. Phys Rev D 46:4304–4319. https://doi.org/10.1103/PhysRevD.46.4304
22. Pretorius F (2005) Evolution of binary black hole spacetimes. Phys Rev Lett 95:121101. arXiv:gr-qc/0507014. https://doi.org/10.1103/PhysRevLett.95.121101
23. Campanelli M, Lousto CO, Marronetti P, Zlochower Y (2006) Accurate evolutions of orbiting black-hole binaries without excision. Phys Rev Lett 96:111101. arXiv:gr-qc/0511048. https://doi.org/10.1103/PhysRevLett.96.111101
24. Baker JG, Centrella J, Choi D-I, Koppitz M, van Meter J (2006) Gravitational wave extraction from an inspiraling configuration of merging black holes. Phys Rev Lett 96:111102. arXiv:gr-qc/0511103. https://doi.org/10.1103/PhysRevLett.96.111102
25. Blanchet L (2002) Gravitational radiation from postNewtonian sources and inspiraling compact binaries. Living Rev Rel 5:3. arXiv:gr-qc/0202016. https://doi.org/10.12942/lrr-2002-3
26. Blanchet L (2006) Gravitational radiation from post-Newtonian sources and inspiralling compact binaries. Living Rev Rel 9:4
27. Will CM (2011) On the unreasonable effectiveness of the post-Newtonian approximation in gravitational physics. Proc Nat Acad Sci 108:5938. arXiv:1102.5192. https://doi.org/10.1073/pnas.1103127108 [gr-qc]
28. Blanchet L (2014) Gravitational radiation from post-newtonian sources and inspiralling compact binaries. Living Rev Rel 17:2. arXiv:1310.1528. https://doi.org/10.12942/lrr-2014-2 [gr-qc]
29. Jaranowski P, Schaefer G (1998) Third postNewtonian higher order ADM Hamilton dynamics for two-body point mass systems. Phys Rev D 57:7274–7291. arXiv:gr-qc/9712075. https://doi.org/10.1103/PhysRevD.57.7274 [Erratum: Phys.Rev.D 63, 029902 (2001)]
30. Damour T, Jaranowski P, Schaefer G (2000) Dynamical invariants for general relativistic two-body systems at the third postNewtonian approximation. Phys Rev D 62:044024. arXiv:gr-qc/9912092. https://doi.org/10.1103/PhysRevD.62.044024
31. Jaranowski P, Schaefer G (1999) The Binary black hole problem at the third postNewtonian approximation in the orbital motion: static part. Phys Rev D 60:124003. arXiv:gr-qc/9906092. https://doi.org/10.1103/PhysRevD.60.124003

32. Damour T, Jaranowski P, Schaefer G (2000) Poincare invariance in the ADM Hamiltonian approach to the general relativistic two-body problem. Phys Rev D 62:021501. arXiv:gr-qc/0003051. https://doi.org/10.1103/PhysRevD.62.021501 [Erratum: Phys.Rev.D 63, 029903 (2001)]
33. Damour T, Jaranowski P, Schaefer G (2001) Equivalence between the ADM-Hamiltonian and the harmonic coordinates approaches to the third postNewtonian dynamics of compact binaries. Phys Rev D 63:044021. arXiv:gr-qc/0010040. https://doi.org/10.1103/PhysRevD.63.044021 [Erratum: Phys.Rev.D 66, 029901 (2002)]
34. Blanchet L, Faye G (2000) Equations of motion of point particle binaries at the third post-Newtonian order. Phys Lett A 271:58. arXiv:gr-qc/0004009. https://doi.org/10.1016/S0375-9601(00)00360-1
35. Blanchet L, Faye G (2001) General relativistic dynamics of compact binaries at the third postNewtonian order. Phys Rev D 63:062005. arXiv:gr-qc/0007051. https://doi.org/10.1103/PhysRevD.63.062005
36. de Andrade VC, Blanchet L, Faye G (2001) Third postNewtonian dynamics of compact binaries: noetherian conserved quantities and equivalence between the harmonic coordinate and ADM Hamiltonian formalisms. Class Quant Grav 18:753–778. arXiv:gr-qc/0011063. https://doi.org/10.1088/0264-9381/18/5/301
37. Damour T, Iyer BR, Sathyaprakash BS (1998) Improved filters for gravitational waves from inspiralling compact binaries. Phys Rev D 57:885–907. arXiv:gr-qc/9708034. https://doi.org/10.1103/PhysRevD.57.885
38. Damour T, Iyer BR, Sathyaprakash BS (2000) Frequency domain P approximant filters for time truncated inspiral gravitational wave signals from compact binaries. Phys Rev D 62:084036. arXiv:gr-qc/0001023. https://doi.org/10.1103/PhysRevD.62.084036
39. Damour T, Iyer BR, Sathyaprakash BS (2001) Detecting binary black holes with efficient and reliable templates. ICTP Lect Notes Ser 3:177–188 arXiv:gr-qc/0012070
40. Buonanno A, Damour T (1999) Effective one-body approach to general relativistic two-body dynamics. Phys Rev D59:084006. arXiv:gr-qc/9811091. https://doi.org/10.1103/PhysRevD.59.084006
41. Buonanno A, Damour T (2000) Transition from inspiral to plunge in binary black hole coalescences. Phys Rev D62:064015. arXiv:gr-qc/0001013. https://doi.org/10.1103/PhysRevD.62.064015
42. Damour T, Jaranowski P, Schaefer G (2000) On the determination of the last stable orbit for circular general relativistic binaries at the third postNewtonian approximation. Phys Rev D62:084011. arXiv:gr-qc/0005034. https://doi.org/10.1103/PhysRevD.62.084011 [gr-qc]
43. Damour T (2016) Gravitational scattering, post-Minkowskian approximation and effective one-body theory. Phys Rev D94:104015. arXiv:1609.00354. https://doi.org/10.1103/PhysRevD.94.104015 [gr-qc]
44. Damour T, Iyer BR, Nagar A (2009) Improved resummation of post-Newtonian multipolar waveforms from circularized compact binaries. Phys Rev D79:064004. arXiv:0811.2069. https://doi.org/10.1103/PhysRevD.79.064004 [gr-qc]
45. Damour T, Nagar A (2016) The effective-one-body approach to the general relativistic two body problem. Lect Notes Phys 905:273–312. https://doi.org/10.1007/978-3-319-19416-5_7
46. Damour T, Nagar A (2007) Faithful effective-one-body waveforms of small-mass-ratio coalescing black-hole binaries. Phys Rev D76:064028. arXiv:0705.2519. https://doi.org/10.1103/PhysRevD.76.064028 [gr-qc]
47. Damour T, Nagar A (2008) Comparing effective-one-body gravitational waveforms to accurate numerical data. Phys Rev D77:024043. arXiv:0711.2628. https://doi.org/10.1103/PhysRevD.77.024043 [gr-qc]
48. Bini D, Damour T (2012) Gravitational radiation reaction along general orbits in the effective one-body formalism. Phys Rev D 86:124012. arXiv:1210.2834. https://doi.org/10.1103/PhysRevD.86.124012 [gr-qc]
49. Khalil M, Buonanno A, Steinhoff A, Vines J (2021) Radiation-reaction force and multipolar waveforms for eccentric, spin-aligned binaries in the effective-one-body formalism. Phys Rev D 104:024046. arXiv:2104.11705. https://doi.org/10.1103/PhysRevD.104.024046 [gr-qc]

50. Maggiore M (2007) Gravitational waves: theory and experiments, vol 1. Oxford University Press. https://doi.org/10.1093/acprof:oso/9780198570745.001.0001
51. Blanchet L (1993) Time asymmetric structure of gravitational radiation. Phys Rev D 47:4392–4420. https://doi.org/10.1103/PhysRevD.47.4392
52. Blanchet L (2005) Gravitational wave tails of tails. Class Quant Grav 15:113–141. arXiv:gr-qc/9710038. https://doi.org/10.1088/0264-9381/15/1/009
53. Mihaylov DP, Ossokine S, Buonanno A, Estelles H, Pompili L, Pürrer M, Ramos-Buades A (2023) pySEOBNR: a software package for the next generation of effective-one-body multipolar waveform models. arXiv:2303.18203 [gr-qc]
54. Khalil M, Buonanno A, Estellés H, Mihaylov DP, Ossokine S, Pompili L, Ramos-Buades A (2023) Theoretical groundwork supporting the precessing-spin two-body dynamics of the effective-one-body waveform models SEOBNRv5. arXiv:2303.18143 [gr-qc]
55. Damour T, Nagar A (2014) New effective-one-body description of coalescing nonprecessing spinning black-hole binaries. Phys Rev D90:044018. arXiv:1406.6913. https://doi.org/10.1103/PhysRevD.90.044018 [gr-qc]
56. Nagar A, Damour T, Reisswig C, Pollney D (2016) Energetics and phasing of nonprecessing spinning coalescing black hole binaries. Phys Rev D93:044046. arXiv:1506.08457. https://doi.org/10.1103/PhysRevD.93.044046 [gr-qc]
57. Nagar A et al (2018) Time-domain effective-one-body gravitational waveforms for coalescing compact binaries with nonprecessing spins, tides and self-spin effects. Phys Rev D 98:104052. arXiv:1806.01772. https://doi.org/10.1103/PhysRevD.98.104052 [gr-qc]
58. Nagar A, Pratten G, Riemenschneider G, Gamba R (2019) A multipolar effective one body model for non-spinning black hole binaries. arXiv:1904.09550 [gr-qc]
59. Nagar A, Riemenschneider G, Pratten G, Rettegno P, Messina F (2020) Multipolar effective one body waveform model for spin-aligned black hole binaries. Phys Rev D 102:024077. arXiv:2001.09082. https://doi.org/10.1103/PhysRevD.102.024077 [gr-qc]
60. Riemenschneider G, Rettegno P, Breschi M, Albertini A, Gamba R, Bernuzzi S, Nagar A (2021) Assessment of consistent next-to-quasicircular corrections and postadiabatic approximation in effective-one-body multipolar waveforms for binary black hole coalescences. Phys Rev D 104:104045. arXiv:2104.07533. https://doi.org/10.1103/PhysRevD.104.104045 [gr-qc]
61. Nagar A, Rettegno P, Gamba R, Albanesi S, Albertini A, Bernuzzi S (2023) Analytic systematics in next generation of effective-one-body gravitational waveform models for future observations. Phys Rev D 108:124018. arXiv:2304.09662. https://doi.org/10.1103/PhysRevD.108.124018 [gr-qc]
62. Berti E et al (2015) Testing general relativity with present and future astrophysical observations. Class Quant Grav 32:243001. arXiv:1501.07274. https://doi.org/10.1088/0264-9381/32/24/243001 [gr-qc]
63. Polchinski, J (2007) String theory: an introduction to the bosonic string. Cambridge Monographs on Mathematical Physics, vol 1. Cambridge University Press. https://doi.org/10.1017/CBO9780511816079
64. Duff MJ (1994) In: The Oskar Klein centenary symposium, pp 22–35. arXiv:hep-th/9410046
65. Clifton T, Ferreira PG, Padilla A, Skordis C (2012) Modified gravity and cosmology. Phys Rep. 513:1–189. arXiv:1106.2476. https://doi.org/10.1016/j.physrep.2012.01.001 [astro-ph.CO]
66. Jordan P (1959) The present state of Dirac's cosmological hypothesis. Z Phys 157:112–121. https://doi.org/10.1007/BF01375155
67. Brans C, Dicke RH (1961) Mach's principle and a relativistic theory of gravitation. Phys Rev 124:925–935. https://doi.org/10.1103/PhysRev.124.925
68. Eardley DM (1975) Observable effects of a scalar gravitational field in a binary pulsar. apjl 196:L59–L62. https://doi.org/10.1086/181744
69. Damour T, Esposito-Farese G (1992) Tensor multiscalar theories of gravitation. Class Quant Grav 9:2093–2176. https://doi.org/10.1088/0264-9381/9/9/015
70. Damour T (1982) Les Houches summer school on gravitational radiation
71. Fierz M (1956) On the physical interpretation of P. Jordan's extended theory of gravitation. Helv Phys Acta 29:128–134

72. Fujii Y, Maeda K-I (2003) The scalar-tensor theory of gravitation. Cambridge monographs on mathematical physics. Cambridge University Press
73. Lang RN (2014) Compact binary systems in scalar-tensor gravity. II. Tensor gravitational waves to second post-Newtonian order. Phys Rev D 89:084014. arXiv:1310.3320. https://doi.org/10.1103/PhysRevD.89.084014 [gr-qc]
74. Lang RN (2015) Compact binary systems in scalar-tensor gravity. III. Scalar waves and energy flux. Phys Rev D 91:084027. arXiv:1411.3073. https://doi.org/10.1103/PhysRevD.91.084027 [gr-qc]
75. Bernard L (2018) Dynamics of compact binary systems in scalar-tensor theories: equations of motion to the third post-Newtonian order. Phys Rev D 98:044004. arXiv:1802.10201. https://doi.org/10.1103/PhysRevD.98.044004 [gr-qc]
76. Bernard L (2019) Dynamics of compact binary systems in scalar-tensor theories: II. Center-of-mass and conserved quantities to 3PN order. Phys Rev D 99:044047. arXiv:1812.04169. https://doi.org/10.1103/PhysRevD.99.044047 [gr-qc]
77. Bernard L (2020) Dipolar tidal effects in scalar-tensor theories. Phys Rev D 101:021501. arXiv:1906.10735. https://doi.org/10.1103/PhysRevD.101.021501 [gr-qc]
78. Doneva DD, Ramazanoğlu FM, Silva HO, Sotiriou TP, Yazadjiev SS (2024) Spontaneous scalarization. Rev Mod Phys 96:015004. arXiv:2211.01766. https://doi.org/10.1103/RevModPhys.96.015004 [gr-qc]
79. Damour T, Esposito-Farese G (1993) Nonperturbative strong field effects in tensor - scalar theories of gravitation. Phys Rev Lett 70:2220–2223. https://doi.org/10.1103/PhysRevLett.70.2220
80. Damour T, Esposito-Farese G (1996) Tensor - scalar gravity and binary pulsar experiments. Phys Rev D 54:1474–1491. arXiv:gr-qc/9602056. https://doi.org/10.1103/PhysRevD.54.1474
81. Freire PC, Wex N, Esposito-Farèse G, Verbiest JP, Bailes M, Jacoby BA, Kramer M, Stairs IH, Antoniadis J, Janssen GH (2012) The relativistic pulsar-white dwarf binary PSR J1738+0333 II. The most stringent test of scalar-tensor gravity. Mon Not Roy Astron Soc 423:3328. arXiv:1205.1450. https://doi.org/10.1111/j.1365-2966.2012.21253.x [astro-ph.GA]
82. Kramer M et al (2021) Strong-field gravity tests with the double pulsar. Phys Rev X 11:041050. arXiv:2112.06795. https://doi.org/10.1103/PhysRevX.11.041050 [astro-ph.HE]
83. Schön O, Doneva DD (2022) Tensor-multiscalar gravity: equations of motion to 2.5 post-Newtonian order. Phys Rev D 105:064034. arXiv:2112.07388. https://doi.org/10.1103/PhysRevD.105.064034 [gr-qc]
84. Sennett N, Marsat S, Buonanno A (2016) Gravitational waveforms in scalar-tensor gravity at 2PN relative order. Phys Rev D 94:084003. arXiv:1607.01420. https://doi.org/10.1103/PhysRevD.94.084003 [gr-qc]
85. Bernard L, Blanchet L, Trestini D (2022) Gravitational waves in scalar-tensor theory to one-and-a-half post-Newtonian order. JCAP 08:(08), 008. arXiv:2201.10924. https://doi.org/10.1088/1475-7516/2022/08/008 [gr-qc]

Part I
Theoretical Waveform Modelling

Chapter 2
Effective-One-Body Hamiltonian in Scalar-Tensor Gravity at the Third Post-Newtonian Order

We determine the general local-in-time effective-one-body (EOB) Hamiltonian for massless Scalar-Tensor (ST) theories at third post-Newtonian (PN) order. For this, we first derive the order-reduced Lagrangian starting from the harmonic coordinate Lagrangian at the third post-Newtonian (PN) order and then obtain its corresponding centre of mass ordinary Hamiltonian. We then map the ordinary Hamiltonian to an effective-one-body (EOB) Hamiltonian at 3PN to derive the EOB metric potentials for the instantaneous part of the dynamics. We also explore the relevance of the 3PN scalar-tensor corrections by studying the innermost stable circular orbit (ISCO).

This is based on the work published in Ref. [1].

2.1 Introduction

Amongst the theories alternative to Einstein's General Relativity (GR), scalar-tensor (ST) theories are most thoroughly studied and tested [2–7]. Besides their manifestly well-posed nature, the presence of an additional gravitational scalar field arises as a natural feature in theories of gravity designed to serve as UV completions of GR. The nonminimally coupled scalar field also gives interesting phenomenology, both in gravitating astrophysical systems and in cosmology [8].

The recent breakthrough in experimental gravity, with the first direct observation of gravitational waves by the LIGO-Virgo Collaboration in 2015 [9], opened new pathways towards probing the dynamics of gravity at extreme conditions [10–16] and led to the first bounds on high-order post-Newtonian coefficients [17]. With the sensitivity of the LIGO [18], Virgo [19] and now KAGRA [20] detector network continuously improving [21], a wealth of detected GW signals emitted by coalescing black-hole (BH) and neutron-star (NS) binaries has been thoroughly analysed to test the strong-field dynamics of GR and probe the nature of the observed compact

T. Jain, *Surfing the Gravitational Waves*, Springer Theses,
https://doi.org/10.1007/978-3-031-96180-9_2

objects [22–25]. So far, in the majority, these are either null-hypothesis tests or searches for generic GR-violating features (dispersion, non-tensorial polarisations, etc.), due to the lack of accurate and complete waveform models in any gravitational theory alternative to GR that are reliable from the early binary inspiral all the way to merger. The beyond GR waveform models are also of interest as deviations from GR could be a systematic error when searching for GW signals, i.e. there could be stealth biases [26] or the signals could be missed altogether. Nevertheless, much progress has been achieved for a selective set of promising theories, including ST, as we shall see in detail below. With this work, we make the next step towards obtaining a model sufficiently accurate for performing direct comparisons between ST and GR. This is already important for reliably interpreting observational bounds using GW data from the current network of detectors and will become even more crucial for probing the strong-field dynamics of gravity to much greater accuracy with the next generation of detectors, such as the Einstein Telescope [27] and Cosmic Explorer [28].

For ST theories in particular, although currently, the most stringent constraints come from binary pulsar observations by studying the effects of scalarized NSs on the orbital evolution [5] there is hope that future gravitational wave (GW) detections of coalescing compact binaries will complement current knowledge by placing additional constraints on the ST parameters using genuine strong-field information from the binary inspiral.

The interpretation of detected GW signals relies on Bayesian data analysis techniques, where the theoretical prediction of a waveform is matched against the data. To this aim, there is an increasing effort in improving the analytical knowledge of the two-body problem in ST theories, both for what concerns the dynamics [29–34] and the waveform generation [35, 36] through post-Newtonian (PN) theory.

To robustly describe the binary dynamics and waveform in the strong field up to the merger, PN results should be recast within the effective-one-body (EOB) description of the two-body problem [37–43]. The generalization of the EOB method to ST theories at the 2PN level has been recently worked out [44, 45]. The aim of this work is to extend the results of Refs. [44, 45] to 3PN order building upon the 3PN Lagrangian in ST of [31, 32].

The Chapter is organised as follows. In Sect. 2.2, we present the conventions used in this thesis for the ST parameters. Following this, in Sect. 2.3 we derive the order-reduced Lagrangian for ST theory at 3PN, and from this, we obtain the centre of mass ordinary Hamiltonian in Sect. 2.4. Finally, in Sect. 2.5, we map the ordinary Hamiltonian into an EOB Hamiltonian at 3PN order and in Sect. 2.6 we explore the relevance of the 3PN ST terms (including nonlocal and finite-size contributions) by studying the ISCO. We use geometric units $c = G = 1$ unless specified explicitly.

2.2 Overview of Scalar-Tensor Parameters

The action of the massless ST theories considering a single scalar field in the Einstein Frame is,

$$S = \frac{c^4}{16\pi G}\int d^4x\sqrt{-g}(R - 2g^{\mu\nu}\partial_\mu\varphi\partial_\nu\varphi) + S_m[\Psi, \mathscr{A}(\varphi)^2 g_{\mu\nu}], \tag{2.1}$$

where $g_{\mu\nu}$ is the Einstein metric, R is the Ricci scalar, φ is the scalar field, Ψ collectively denotes the matter fields, $g \equiv \det(g_{\mu\nu})$ and G is the Newtonian constant.

When considering compact, self-gravitating objects in ST theories, we also need to consider the size and internal gravity of each body. Therefore, following the approach pioneered by Eardley in Ref. [46] which reflects that the gravitational binding energy of the body depends on the value of the gravitational constant, we consider that the total mass of each body is described by the mass function dependent on the value of the scalar field (φ) in an undefined manner at the location of the particles. This is because the effective gravitational constant in ST theories depends on the scalar field.

Accordingly, the matter action in ST theories will have an effective dependence on the scalar field φ, i.e.

$$S_m = -\sum_{I=A,B}\int\sqrt{-\tilde{g}_{\mu\nu}\frac{dx^\mu}{d\lambda}\frac{dx^\nu}{d\lambda}}\tilde{m}_I(\varphi), \tag{2.2}$$

where λ is an affine parameter, $\tilde{g}_{\mu\nu}$ is the Jordan frame metric and $\tilde{m}_I(\varphi)$ is the Jordan-frame mass of body I. As a result of this, the field equations and hence the dynamics of the system will depend on the variation in the mass of each body ("sensitivity") for the change in the scalar field.

In this Chapter, we will work in the Einstein frame, therefore using the conformal transformation,

$$\tilde{g}_{\mu\nu} = \mathscr{A}(\varphi)^2 g_{\mu\nu}, \tag{2.3}$$

the Einstein frame mass $m(\varphi)$ is defined as

$$m(\varphi) = \mathscr{A}(\varphi)\tilde{m}(\varphi). \tag{2.4}$$

Here, $g_{\mu\nu}$ is the Einstein-frame metric and $\mathscr{A}(\varphi)$ is called the coupling function. The ST theory is uniquely fixed when the function $\mathscr{A}(\varphi)$ is defined, and when $\mathscr{A}(\varphi) = \text{const.}$, GR is recovered. The dynamics of the scalar field originates from its coupling to the matter fields in the Einstein frame and is measured by the parameter,

$$\alpha(\varphi) = \frac{d\ln\mathscr{A}(\varphi)}{d\varphi}. \tag{2.5}$$

Table 2.1 Relation between the ST parameters used in the two-body Lagrangian of Ref. [31], the DEF ones and the slightly simplified notation that we are using here. The index "0" signifies a quantity evaluated at $\varphi = \varphi_0$, where φ_0 is the asymptotic constant value of the scalar field far from the two bodies

LB [31]	DEF [2, 4]	This thesis
m_1	$m_A^0/\mathscr{A}_0$	$m_A^0/\mathscr{A}_0 \equiv \tilde{m}_A^0$
m_2	$m_B^0/\mathscr{A}_0$	$m_B^0/\mathscr{A}_0 \equiv \tilde{m}_B^0$
α	$\frac{1+\alpha_A^0\alpha_B^0}{1+\alpha_0^2}$	α_{AB}
$\tilde{G}\alpha$	$(1+\alpha_A^0\alpha_B^0)\mathscr{A}_0^2 \equiv G_{AB}\mathscr{A}_0^2$	$G_{AB}\mathscr{A}_0^2 \equiv \tilde{G}_{AB}$
$\bar{\gamma}$	$-2\frac{\alpha_A^0\alpha_B^0}{1+\alpha_A^0\alpha_B^0} \equiv \bar{\gamma}_{AB}$	$\bar{\gamma}_{AB}$
$\bar{\beta}_1$	$\frac{1}{2}\frac{(\beta_A\alpha_B^2)_0}{(1+\alpha_A^0\alpha_B^0)^2} \equiv \bar{\beta}^A_{BB}$	$\bar{\beta}_A$
$\bar{\beta}_2$	$\frac{1}{2}\frac{(\beta_B\alpha_A^2)_0}{(1+\alpha_A^0\alpha_B^0)^2} \equiv \bar{\beta}^B_{AA}$	$\bar{\beta}_B$
$\bar{\delta}_1$	$\frac{(\alpha_A^0)^2}{(1+\alpha_A^0\alpha_B^0)^2}$	δ_A
$\bar{\delta}_2$	$\frac{(\alpha_B^0)^2}{(1+\alpha_A^0\alpha_B^0)^2}$	δ_B
$\bar{\chi}_1$	$-\frac{1}{4}\frac{(\beta'_A\alpha_B^3)_0}{(1+\alpha_A^0\alpha_B^0)^3} \equiv -\frac{1}{4}\varepsilon^A_{BBB}$	$-\frac{1}{4}\varepsilon_A$
$\bar{\chi}_2$	$-\frac{1}{4}\frac{(\beta'_B\alpha_A^3)_0}{(1+\alpha_A^0\alpha_B^0)^3} \equiv -\frac{1}{4}\varepsilon^B_{AAA}$	$-\frac{1}{4}\varepsilon_B$
$\bar{\beta}_1\bar{\beta}_2/\bar{\gamma}$	$-\frac{1}{8}\frac{\beta_A^0\alpha_A^0\beta_B^0\alpha_B^0}{(1+\alpha_A^0\alpha_B^0)^3} \equiv -\frac{1}{8}\zeta_{ABAB}$	$-\frac{1}{8}\zeta$
$\bar{\kappa}_1$	$\frac{(\alpha_B^4\beta''_A)_0}{8(1+\alpha_A^0\alpha_B^0)^4}$	$\bar{\kappa}_A$
$\bar{\kappa}_2$	$\frac{(\alpha_A^4\beta''_B)_0}{8(1+\alpha_A^0\alpha_B^0)^4}$	$\bar{\kappa}_B$

The dimensionless body-dependent parameters that encompass the scalar-field effect in the Einstein frame up to 3PN order can be defined using the Einstein frame mass functions $m_I(\varphi)$ following Refs. [2, 4, 45] i.e.

$$\alpha_I = \frac{d\ln m(\varphi)_I}{d\varphi}, \tag{2.6}$$

$$\beta_I = \frac{d\alpha_I}{d\varphi}, \tag{2.7}$$

$$\beta'_I = \frac{d\beta_I}{d\varphi}, \tag{2.8}$$

$$\beta''_I = \frac{d\beta'_I}{d\varphi}. \tag{2.9}$$

The ST parameters entering the two-body Lagrangian at 3PN of Ref. [31] are defined in the Jordan frame. Since we work in the Einstein frame in this Chapter (Part I of the thesis), we convert these Jordan-frame parameters into the Einstein-frame ones, i.e. Damour-Esposito-Farese (DEF) conventions, and the corresponding parameters used in this Chapter using Eqs. (2.3) and (2.4). The relation of the ST parameters between the Jordan frame and the Einstein frame is given in Table 2.1.

2.3 Scalar-Tensor 3PN Ordinary Lagrangian

The ST two-body 3PN Lagrangian was obtained in Ref. [32] in harmonic coordinates such that it depends (linearly) on the acceleration of the two bodies. In this Section, we transform the Lagrangian of Ref. [32] into an ordinary Lagrangian, that is the Lagrangian dependent only on the positions and the velocities of the two bodies. At 2PN order this can be achieved, either by boldly replacing the acceleration by the equations of motion or by using the 4-dimensional coordinate transformation in position variables of particle so called the contact transformation (modulo a total time derivative which is irrelevant in a Lagrangian) [47]. Here, for constructing a generic transformation at 3PN order, we will choose the second approach, i.e. we will use the contact transformation to make a 4-dimensional coordinate change in position variables of particle to eliminate the acceleration dependence in the Lagrangian.

The order-reduced Lagrangian obtained using the contact transformation is

$$\mathscr{L}_f^{\mathrm{red}'} = \mathscr{L} + \sum_A \delta Z_A^i \frac{\delta \mathscr{L}}{\delta Z_A^i}, \tag{2.10}$$

where $\mathscr{L}$ is the Lagrangian in the harmonic coordinate system, δZ_A is the 4-dimensional coordinate change, $\frac{\delta \mathscr{L}}{\delta Z_I^i}$ is the functional derivative term defined as

$$\frac{\delta \mathscr{L}}{\delta Z_I^i} \equiv \frac{\partial \mathscr{L}}{\partial Z_I^i} - \frac{d}{dt}\left(\frac{\partial \mathscr{L}}{\partial V_I^i}\right) + \frac{d^2}{dt^2}\left(\frac{\partial \mathscr{L}}{\partial A_I^i}\right) = 0. \tag{2.11}$$

Here, Z_I^i, $V_I^i = dZ_I^i/dt$ and $A_I^i = dV_I^i/dt$ are the position, velocity and acceleration, respectively.

To construct a general contact transformation, we can also freely add a term of the type $\partial_{V_A^i} \mathscr{F}/m_A^0$ to the 4-dimensional coordinate change δZ_A (see, Ref. [48] for GR), where $\mathscr{F}$ is an arbitrary function that depends only on the positions and the velocities. The total time derivative of $\mathscr{F}$ then eliminates the acceleration produced by the addition of this term, without affecting the dynamics of the system. Therefore, the general ordinary (reduced) Lagrangian, $\mathscr{L}_f^{\mathrm{red}}$, is given by

$$\mathscr{L}_f^{\mathrm{red}} = \mathscr{L} + \frac{d\mathscr{F}}{dt} + \sum_A \delta Z_A^i \frac{\delta \mathscr{L}}{\delta Z_A^i}. \tag{2.12}$$

As the dependence of the Lagrangian on the accelerations starts at 2PN order, the 4-dimensional coordinate change (contact transformation) δZ_A and hence, the arbitrary function $\mathscr{F}$ also start at 2PN order. Therefore, only the leading order Newtonian contribution of the fractional derivative term, $\frac{\delta \mathscr{L}}{\delta Z_I^i}$ is required in Eq. (2.10) to construct the contact transformation at 2PN order. Now, to construct the transformation at 3PN order, the fractional derivative term should be considered up to 1PN order, i.e.,

$$\frac{\delta \mathscr{L}}{\delta Z_I^i} = m_I^0 \left(-a_I^i - \sum_{I \neq J} \frac{G_{AB} m_J}{R^2} n_{AB}^i + \frac{1}{c^2} C_I^i \right), \tag{2.13}$$

where C_I^i is the 1PN contribution which can be derived by using up to 1PN order harmonic-coordinate Lagrangian given in Ref. [32] in Eq. (2.11).

The 1PN term C_I^i in the above equation depends on accelerations, so we will get additional acceleration-dependent terms (on multiplying with the 2PN contribution of the contact transformation δZ_I^i) in the reduced Lagrangian at 3PN order which do not cancel with the acceleration-dependent terms of the harmonic coordinate Lagrangian. Hence, the 2PN method presented in Ref. [45] for ST theory cannot be extended at 3PN order. In contrast, following Ref. [49] for the GR case, we have to introduce some sort of 'counter term' at 3PN order for ST theory in order to eliminate the additional acceleration-dependent terms entering due to the 1PN contribution C_I^i. This method to construct a contact transformation at 3PN order is based on the algorithm presented in Ref. [50] to eliminate higher-order derivative terms from the Lagrangian. After introducing the 'counter term' X_I^i, the contact transformation at 3PN order becomes, [49],

$$\delta Z_I^i = \frac{1}{m_I^0} \left(q_A^i + \frac{\partial \mathscr{F}}{\partial V_I^i} + X_I^i \right), \tag{2.14}$$

where X_I^i is defined as (see, Eq. (3.17) of Ref. [49]),

$$X_I^i = \sum_J \left[q_J^j + \frac{\partial \mathscr{F}}{\partial V_J^j} \right] \frac{\partial C_J^j}{\partial A_I^i}, \tag{2.15}$$

and $q_I^i = \frac{\partial \mathscr{L}}{\partial a_I^i}$ are the conjugate momenta of acceleration.

As mentioned previously, $\mathscr{F}$ starts at 2PN order, therefore, up to 3PN order it can be formally written as

$$\mathscr{F} = \mathscr{F}_{\mathrm{2PN}} + \mathscr{F}_{\mathrm{3PN}}, \tag{2.16}$$

where $\mathscr{F}_{\mathrm{2PN}}$ is the 2PN contribution given in Ref. [45], and $\mathscr{F}_{\mathrm{3PN}}$ is the 3PN contribution, the most general form of which is given in Eq. (2.17). This 3PN contribution depends on 56 parameters, and the effective Newton's gravitational constant G_{AB} appears in the definition of parameters F_i of $\mathscr{F}_{\mathrm{3PN}}$ for dimensional convenience. We first derive the contact transformation at 3PN order for ST theory (see Appendix 2.8), and then use it to derive the reduced Lagrangian, $\mathscr{L}_f^{\mathrm{red}}$, at 3PN order. This reduced Lagrangian is an ordinary Lagrangian, dependent on F_i and f_i.[1]

Hence, we have a whole class of coordinate systems (dependent on the parameters F_i and f_i) in which our derived Lagrangian $\mathscr{L}_f^{\mathrm{red}}$ is ordinary, while harmonic coordinates do not belong to this class.

[1] By f_i, we denote the parameters of the generic function at 2PN order given in Ref. [45].

For completeness, here we give the explicit expression of $\mathcal{F}^{3\mathrm{PN}}$, that reads

$$
\begin{aligned}
\mathcal{F}_{3\mathrm{PN}} =& G_{AB} m_A^0 m_B^0 \Big\{ \Big[F_1 (\vec{V}_A \cdot \vec{V}_A)^2 + F_2 (\vec{V}_A \cdot \vec{V}_A) \vec{V}_A \cdot \vec{V}_B + F_3 (\vec{V}_A \cdot \vec{V}_B)^2 + F_4 (\vec{V}_A \cdot \vec{V}_A)(\vec{V}_B \cdot \vec{V}_B) \\
&+ F_5 (\vec{V}_B \cdot \vec{V}_B) \vec{V}_A \cdot \vec{V}_B + F_6 (\vec{V}_B \cdot \vec{V}_B)^2 \Big] \vec{n}_{AB} \cdot \vec{V}_A - \Big[F_7 (\vec{V}_A \cdot \vec{V}_A)^2 + F_8 (\vec{V}_A \cdot \vec{V}_A) \vec{V}_A \cdot \vec{V}_B \\
&+ F_9 (\vec{V}_A \cdot \vec{V}_A)(\vec{V}_B \cdot \vec{V}_B) + F_{10} (\vec{V}_A \cdot \vec{V}_B)^2 + F_{11} (\vec{V}_B \cdot \vec{V}_B) \vec{V}_A \cdot \vec{V}_B + F_{12} (\vec{V}_B \cdot \vec{V}_B)^2 \Big] \vec{n}_{AB} \cdot \vec{V}_B + [F_{13} \vec{V}_A \cdot \vec{V}_A \\
&+ F_{14} \vec{V}_A \cdot \vec{V}_B + F_{15} \vec{V}_B \cdot \vec{V}_B] (\vec{n}_{AB} \cdot \vec{V}_A)^3 + [F_{16} \vec{V}_A \cdot \vec{V}_A + F_{17} \vec{V}_A \cdot \vec{V}_B + F_{18} \vec{V}_B \cdot \vec{V}_B] (\vec{n}_{AB} \cdot \vec{V}_A)^2 (\vec{n}_{AB} \cdot \vec{V}_B) \\
&- [F_{19} \vec{V}_A \cdot \vec{V}_A + F_{20} \vec{V}_A \cdot \vec{V}_B + F_{21} \vec{V}_B \cdot \vec{V}_B] (\vec{n}_{AB} \cdot \vec{V}_A)(\vec{n}_{AB} \cdot \vec{V}_B)^2 \\
&- [F_{22} \vec{V}_A \cdot \vec{V}_A + F_{23} \vec{V}_A \cdot \vec{V}_B + F_{24} \vec{V}_B \cdot \vec{V}_B] (\vec{n}_{AB} \cdot \vec{V}_B)^3 + F_{25} (\vec{n}_{AB} \cdot \vec{V}_A)^5 + F_{26} (\vec{n}_{AB} \cdot \vec{V}_A)^4 \vec{n}_{AB} \cdot \vec{V}_B \\
&+ F_{27} (\vec{n}_{AB} \cdot \vec{V}_A)^3 (\vec{n}_{AB} \cdot \vec{V}_B)^2 - F_{28} (\vec{n}_{AB} \cdot \vec{V}_A)^2 (\vec{n}_{AB} \cdot \vec{V}_B)^3 - F_{29} (\vec{n}_{AB} \cdot \vec{V}_A)(\vec{n}_{AB} \cdot \vec{V}_B)^4 - F_{30} (\vec{n}_{AB} \cdot \vec{V}_B)^5 \Big\} \\
&+ \frac{G_{AB}^2 (m_A^0)^2 m_B^0}{R} \Big\{ \Big[F_{31} \vec{V}_A \cdot \vec{V}_A + F_{32} \vec{V}_A \cdot \vec{V}_B + F_{33} \vec{V}_B \cdot \vec{V}_B \Big] \vec{n}_{AB} \cdot \vec{V}_A + F_{37} (\vec{n}_{AB} \cdot \vec{V}_A)^3 \\
&+ F_{38} \vec{n}_{AB} \cdot \vec{V}_B (\vec{n}_{AB} \cdot \vec{V}_A)^2 - \Big[F_{34} \vec{V}_A \cdot \vec{V}_A + F_{35} \vec{V}_A \cdot \vec{V}_B + F_{36} \vec{V}_B \cdot \vec{V}_B \Big] \vec{n}_{AB} \cdot \vec{V}_B - F_{39} (\vec{n}_{AB} \cdot \vec{V}_B)^2 (\vec{n}_{AB} \cdot \vec{V}_A) \\
&- F_{40} (\vec{n}_{AB} \cdot \vec{V}_B)^3 \Big\} \\
&+ \frac{G_{AB}^2 m_A^0 (m_B^0)^2}{R} \Big\{ (F_{41} \vec{V}_A \cdot \vec{V}_A + F_{42} \vec{V}_A \cdot \vec{V}_B + F_{43} \vec{V}_B \cdot \vec{V}_B) \vec{n}_{AB} \cdot \vec{V}_A - (F_{44} \vec{V}_A \cdot \vec{V}_A + F_{45} \vec{V}_A \cdot \vec{V}_B \\
&+ F_{46} \vec{V}_B \cdot \vec{V}_B) \vec{n}_{AB} \cdot \vec{V}_B + F_{47} (\vec{n}_{AB} \cdot \vec{V}_A)^3 + F_{48} \vec{n}_{AB} \cdot \vec{V}_B (\vec{n}_{AB} \cdot \vec{V}_A)^2 - F_{49} \vec{n}_{AB} \cdot \vec{V}_A (\vec{n}_{AB} \cdot \vec{V}_B)^2 \\
&- F_{50} (\vec{n}_{AB} \cdot \vec{V}_B)^3 \Big\} \\
&+ \frac{G_{AB}^3 (m_A^0)^2 (m_B^0)^2}{R^2} \Big\{ F_{51} \vec{n}_{AB} \cdot \vec{V}_A - F_{52} \vec{n}_{AB} \cdot \vec{V}_B \Big\} + \frac{G_{AB}^3 (m_A^0)^3 m_B^0}{R^2} \Big\{ F_{53} \vec{n}_{AB} \cdot \vec{V}_A - F_{54} \vec{n}_{AB} \cdot \vec{V}_B \Big\} \\
&+ \frac{G_{AB}^3 m_A^0 (m_B^0)^3}{R^2} \Big\{ F_{55} \vec{n}_{AB} \cdot \vec{V}_A - F_{56} \vec{n}_{AB} \cdot \vec{V}_B \Big\} ,
\end{aligned}
\tag{2.17}
$$

where G_{AB} is the effective Newton's gravitational constant in DEF conventions (see Table 2.1) [2, 4], $\vec{V}_{A,B}$ are the velocities of the two bodies, $\vec{n}_{AB} \equiv (\vec{Z}_A - \vec{Z}_B)/R$ the unit vector of the relative separation and $\vec{Z}_{A,B}$ indicates the positions of the two bodies with radial separation $R \equiv |\vec{Z}_A - \vec{Z}_B|$.

2.4 Center of Mass Frame Two-Body Ordinary Hamiltonian at 3PN

Let us now derive the ordinary Hamiltonian, in the centre of mass (COM) frame, corresponding to the class of ordinary Lagrangians of Sect. 2.3. We do so by ordinary Legendre transformation. In the COM frame, the total momentum vanishes, $\vec{P}_A + \vec{P}_B = 0$, where $\vec{P}_{A,B}$ are the momenta of the two bodies, so the conjugate variables are $\vec{Z} = \vec{Z}_A - \vec{Z}_B$ and $\vec{P} = \vec{P}_A = -\vec{P}_B$. Since we are considering non-spinning bodies, the motion is planar and we use polar coordinates (R, ϕ) with conjugate momenta (P_R, P_ϕ), setting $\theta = \pi/2$. The general structure of the isotropic, time-translation invariant Hamiltonian at 2PN in the COM frame is presented in Eqs. (III.15)–(III.16) of Ref. [45]. By defining $M \equiv m_A^0 + m_B^0$ the total mass of the system, and $\mu \equiv m_A^0 m_B^0 / M$ its reduced mass, we can define the convenient to use mass reduced variables (always indicated with hat superscript, here and below)

$(\hat{P}, \hat{P}_R, \hat{R})$, where $\hat{P}^2 \equiv P^2/\mu^2 = \hat{P}_R^2 + \hat{P}_\phi^2/\hat{R}^2$, $\hat{P}_R \equiv P_R/\mu$, $\hat{P}_\phi \equiv P_\phi/(\mu M)$ and $\hat{R} \equiv R/M$, so that the 3PN contribution formally reads

$$\hat{H}^{3\text{PN}} = \left(h_1^{3\text{PN}} \hat{P}^8 + h_2^{3\text{PN}} \hat{P}^6 \hat{P}_R^2 + h_3^{3\text{PN}} \hat{P}^4 \hat{P}_R^4 + h_4^{3\text{PN}} \hat{P}^2 \hat{P}_R^6 + h_5^{3\text{PN}} \hat{P}_R^8\right) + \frac{1}{\hat{R}} \left(h_6^{3\text{PN}} \hat{P}^6 + h_7^{3\text{PN}} \hat{P}^4 \hat{P}_R^2 + h_8^{3\text{PN}} \hat{P}^2 \hat{P}_R^4 + h_9^{3\text{PN}} \hat{P}_R^6\right) + \frac{1}{\hat{R}^2} \left(h_{10}^{3\text{PN}} \hat{P}^4 + h_{11}^{3\text{PN}} \hat{P}^2 \hat{P}_R^2 + h_{12}^{3\text{PN}} \hat{P}_R^4\right) + \frac{1}{\hat{R}^3} \left(h_{13}^{3\text{PN}} \hat{P}^2 + h_{14}^{3\text{PN}} \hat{P}_R^2\right) + \frac{1}{\hat{R}^4} h_{15}^{3\text{PN}}. \tag{2.18}$$

Here, the $h_i^{3\text{PN}}$ terms formally indicate the numerical coefficients that we are going to calculate. Before doing so explicitly and introducing our results, let us recall an important technical fact. The ordinary Hamiltonian directly obtained from the class of Lagrangians of Sect. 2.3 via Legendre transformation contains two undetermined constants r_1' and r_2'. These constants parametrise logarithmic terms and are directly inherited from the harmonic coordinates Lagrangian of Ref. [32], where they arise due to the regularisation procedure (via Hadamard partie finie technique). However, Ref. [51] found that these constants are absent in the 3PN ordinary Hamiltonian in GR. As shown in Ref. [52], the reason why this is so is that the two constants, and the related logarithmic terms, can be gauged away from the harmonic coordinate Lagrangian, in accordance with the fact that these are pure gauge quantities. More precisely, Ref. [49] showed that, in the GR case, the regularization constants can also be eliminated by including a logarithmic dependence in the function $\mathscr{F}$ of the contact transformation. Therefore, to gauge away the dependence on r_1' and r_2' from the ordinary Hamiltonian of ST theory, we take an analogy with the approach of Ref. [49] in GR and add the most generic logarithmic dependent terms,

$$\frac{G_{AB}^3 (m_A^0)^3 m_B^0}{R^2} \left(F_{57} \log \frac{r_1'}{R}\right) + \frac{G_{AB}^3 m_A^0 (m_B^0)^3}{R^2} \left(F_{58} \log \frac{r_2'}{R}\right), \tag{2.19}$$

to our arbitrary function $\mathscr{F}_{3PN}$. In ST theory, the coefficients F_{57} and F_{58} must be

$$F_{57} = \left[-\frac{11}{4}(2 + \bar{\gamma}_{AB})^2 + \delta_A\right], \tag{2.20}$$

$$F_{58} = \left[-\frac{11}{4}(2 + \bar{\gamma}_{AB})^2 + \delta_B\right], \tag{2.21}$$

in order to remove the gauge dependence on r_1' and r_2' from the ordinary class of Hamiltonians. The so-obtained complete expression of the ordinary Hamiltonian coefficients of Eq. (2.18) at 3PN order is given in Appendix 2.9.

2.5 Scalar-Tensor Deformation of the 3PN Effective-One-Body Hamiltonian

Let us now turn to discuss the main result of this Chapter, i.e. the instantaneous 3PN contribution to the EOB potentials in ST theory. The mapping between the two-body ordinary (Arnowitt-Deser-Misner (ADM))-like)[2] Hamiltonian and the EOB Hamiltonian can be done using different procedures (e.g., Delaunay Hamiltonian, canonical transformation, comparison of the periastron advance, see e.g. Refs. [37, 39, 53, 54]). Here we will use the canonical transformation approach, adapting the procedure of Ref. [39].

2.5.1 *Canonical Transformation at 3PN Order*

To fix notation, we indicate with (Q, P) the conjugate variables of the real, ordinary, two-body Hamiltonian, while with (q, p) those of the EOB Hamiltonian. We start from the 2PN-accurate canonical transformation in ST theory of Ref. [45] (see Eqs. (III.23) and (III.25) therein) and augment it with 3PN terms. We do so by modifying the generating function $G(Q, p)^{3\text{PN}}$ of the canonical transformation up to 3PN order, whose 3PN contribution reads

$$\frac{G(Q, p)^{3\text{PN}}}{\mu M} = \hat{R}\,\hat{p}_r\bigg[\left(\alpha_3\mathscr{P}^6 + \beta_3\mathscr{P}^4\hat{p}_r^2 + \gamma_3\mathscr{P}^2\hat{p}_r^4 + \delta_3\hat{p}_r^6\right) + \frac{1}{\hat{R}}(\varepsilon_3\mathscr{P}^4 + \eta_3\mathscr{P}^2\hat{p}_r^2 + \theta_3\hat{p}_r^4) + \frac{1}{\hat{R}^2}(\lambda_3\mathscr{P}^2 + \rho_3\hat{p}_r^2) + \frac{\sigma_3}{\hat{R}^3}\bigg], \tag{2.22}$$

where $\mathscr{P}$, $\hat{p}_r$, and $\hat{R}$ are the dimensionless variables defined as [45],

$$\mathscr{P}^2 \equiv \hat{p}_r^2 + \frac{\hat{p}_\phi^2}{\hat{R}^2}, \quad \hat{R} \equiv \frac{R}{M}, \quad \hat{p}_r^2 \equiv \frac{p_r}{\mu}, \quad \hat{p}_\phi \equiv \frac{p_\phi}{\mu M}, \tag{2.23}$$

and $(\alpha_3, \beta_3, \ldots, \sigma_3)$ formally indicate the ten 3PN coefficients.

We will then follow the same procedure as Ref. [45], i.e. we will express the real and effective Hamiltonian in an intermediate coordinate system (Q, p) in order to match the two.

[2] Here, we use ADM-like to refer to the ordinary coordinates as the ADM coordinates fall under the special category of ordinate coordinate systems in which the Lagrangian is not dependent on the acceleration and its higher derivatives (see, Ref. [48]).

2.5.2 Scalar-Tensor Effective-One-Body Hamiltonian at 3PN: Instantaneous Part

Within the EOB approach, the real EOB Hamiltonian is related to the effective Hamiltonian by

$$\hat{H}_{\text{real}} \equiv \frac{H_{\text{real}}}{\mu} = \frac{1}{\nu}\sqrt{1 + 2\nu\left(\hat{H}_{\text{eff}} - 1\right)}, \tag{2.24}$$

where $\nu = \mu/M$ is the symmetric mass ratio and $\hat{H}_{\text{eff}} \equiv H_{\text{eff}}/\mu$ is the reduced-mass effective Hamiltonian. This relation was originally proved to be correct up to 3PN order in GR [39]. Recently, within the Post-Minkowskian scheme, Ref. [55] proved it to hold at *all* PN orders, both in GR and in ST theories. Here we choose to incorporate the 3PN terms within the EOB Hamiltonian following the scheme of Ref. [39], i.e. by writing the effective Hamiltonian $\hat{H}_{\text{eff}}$ as

$$\hat{H}_{\text{eff}} = \sqrt{A(\hat{r})\left[1 + \frac{\hat{p}_\phi^2}{\hat{r}^2} + \frac{\hat{p}_r^2}{B(\hat{r})} + \hat{Q}_e(\hat{r}, \hat{p})\right]}\,. \tag{2.25}$$

Here, we indicate with $(\phi, \hat{p}_\phi, \hat{r}, \hat{p}_r)$ the canonical variables, with $\hat{p}^2 = \hat{p}_r^2 + \hat{p}_\phi^2/\hat{r}^2$, while (A, B, Q_e) are the EOB potentials. The structure of the nongeodesic term $\hat{Q}_e \equiv Q_e/\mu^2$ at 3PN order reads

$$\hat{Q}_e(\hat{r}, \hat{p}) = \frac{1}{\hat{r}^2}\left(q_1 \hat{p}^4 + q_2 \hat{p}^2 \hat{p}_r^2 + q_3 \hat{p}_r^4\right). \tag{2.26}$$

Following Ref. [39], we use the gauge freedom at our disposal to set $q_1 = q_2 = 0$, so that the Q_e function only depends on the radial momentum. This choice is known as the Damour-Jaranowski-Schäfer (DJS) gauge, first introduced in Ref. [39]. We recall, in passing, that this is just one among the many (actually infinite) possibilities of devising an effective dynamics based on a generalized mass shell condition $g_{\text{eff}}^{\mu\nu} p_\mu p_\nu + Q_e(p) = 0$ and the relation between the effective and real Hamiltonian given by Eq. (2.24) [55–61]. In the DJS gauge, the three EOB potentials at 3PN order formally read

$$A(\hat{r}) = 1 - \frac{2}{\hat{r}} + \frac{a_2}{\hat{r}^2} + \frac{a_3}{\hat{r}^3} + \frac{a_4}{\hat{r}^4}, \tag{2.27}$$

$$B(\hat{r}) = 1 + \frac{b_1}{\hat{r}} + \frac{b_2}{\hat{r}^2} + \frac{b_3}{\hat{r}^3}, \tag{2.28}$$

$$\hat{Q}_e(\hat{r}, \hat{p}) = \frac{q_3}{\hat{r}^2}\hat{p}_r^4, \tag{2.29}$$

where the a_i and b_i terms are the symmetric mass ratio ($\nu = \mu/M$)-dependent deformations of the Schwarzschild metric potentials, that take into account both GR and

ST corrections.[3] The GR and ST contributions are separated as

$$A = A^{\mathrm{GR}} + \delta A^{\mathrm{ST}}, \tag{2.30}$$

$$B = B^{\mathrm{GR}} + \delta B^{\mathrm{ST}}, \tag{2.31}$$

$$\hat{Q}_e = \hat{Q}_e^{\mathrm{GR}} + \delta \hat{Q}_e^{\mathrm{ST}}, \tag{2.32}$$

which reflects in the ν-dependent contributions as

$$a_i = a_i(\nu) = a_i^{\mathrm{GR}}(\nu) + \delta a_i^{\mathrm{ST}}(\nu), \tag{2.33}$$

$$b_i = b_i(\nu) = b_i^{\mathrm{GR}}(\nu) + \delta b_i^{\mathrm{ST}}(\nu), \tag{2.34}$$

$$q_3 = q_3(\nu) = q_3^{\mathrm{GR}}(\nu) + \delta q_3^{\mathrm{ST}}(\nu). \tag{2.35}$$

The GR terms are known analytically up to 6PN order [62, 63], except for some yet unknown coefficients proportional to ν^2. As mentioned above, the 2PN ST corrections to the EOB potentials have been computed in Refs. [44, 45].

Before proceeding with the computations, let us recall the technical fact that starting from the 3PN order in ST theory, nonlocal-in-time and finite-size effect contributions have also to be added to the local terms. In this Chapter, we will focus on these latter contributions and we postpone the computation of the complete nonlocal terms to Chap. 3; see, Sect. 2.5.3 for the nonlocal terms restricted to the circular case.

Let us now compute the local ST contributions to the EOB potentials at 3PN order, i.e. $(\delta a_{4,\mathrm{loc}}^{\mathrm{ST}}, \delta b_{3,\mathrm{loc}}^{\mathrm{ST}}, \delta q_{3,\mathrm{loc}}^{\mathrm{ST}})$. We do so by first applying canonical transformation of Sect. 2.5.1 to both the real two-body ordinary Hamiltonian $\hat{H}_{\mathrm{real}}$ of Sect. 2.4 and the 3PN EOB Hamiltonian ($\hat{H}_{\mathrm{eff}}$) to express them in the intermediate coordinate system (Q, p). Then, the canonically transformed $\hat{H}_{\mathrm{real}}$ is matched with the PN expanded $\hat{H}_{\mathrm{real}}$ obtained using Eq. (2.24) to find the unique solution of a_i, b_i, and q_i up to 3PN order. For the new 3PN-order ST contributions, we get:

$$\begin{aligned}
\delta a_{4,\mathrm{loc}}^{\mathrm{ST}} = & -2\bar{\gamma}_{AB} - \frac{13}{2}\bar{\gamma}_{AB}^2 - 5\bar{\gamma}_{AB}^3 + \left[-\frac{7}{6}(2+\bar{\gamma}_{AB})^2 + 9\bar{\gamma}_{AB}^2\right]\langle\bar{\beta}\rangle - \bar{\gamma}_{AB}\langle\varepsilon\rangle + \frac{2}{3}\langle\bar{\kappa}\rangle + \frac{2}{3}(1+2\bar{\gamma}_{AB})\langle\delta\rangle \\
& + 6(\beta_+^2 + \beta_-^2) - 12X_{AB}\beta_-\beta_+ + \frac{2}{3}(\beta_-(\delta_- + X_{AB}\delta_+) - \langle\delta\rangle\beta_+) + \nu\Bigg\{\frac{11}{4\alpha_{AB}}(2+\bar{\gamma}_{AB})\bar{\gamma}_{AB} \\
& - \frac{\langle\delta\rangle}{\alpha_{AB}(2+\bar{\gamma}_{AB})} + \frac{32}{\bar{\gamma}_{AB}^2}\langle\bar{\beta}\rangle(\beta_+^2 - \beta_-^2) + \frac{4}{\bar{\gamma}_{AB}}\Bigg[\beta_-\left(\varepsilon_- - X_{AB}\varepsilon_+ + \frac{4}{3}(2\delta_- - X_{AB}\delta_+)\right) \\
& + \beta_+\left(\frac{4}{3}(2\delta_+ - X_{AB}\delta_-) - \langle\varepsilon\rangle\right)\Bigg] + \bar{\gamma}_{AB}\Bigg[\frac{581}{18} - \frac{75}{64}\pi^2 - 8\zeta - \frac{32}{3}\langle\bar{\beta}\rangle - 20\beta_+ \\
& + \frac{1}{2}\langle\delta\rangle + \left(\frac{4}{3} + \frac{7}{32}\pi^2\right)\delta_+ + 2\varepsilon_+\Bigg] + \bar{\gamma}_{AB}^2\left(\frac{239}{18} - \frac{5}{32}\pi^2 - \frac{2}{3}\langle\bar{\beta}\rangle\right) \\
& + \bar{\gamma}_{AB}^3\left(-\frac{3}{8} + \frac{7}{128}\pi^2\right) + 3\zeta - 6(\beta_+^2 + \beta_-^2) - \frac{5}{3}\langle\bar{\beta}\rangle + \frac{3}{2}X_{AB}\beta_- - 12X_{AB}\beta_-\beta_+ \\
& - \frac{8}{3}(\delta_+\langle\bar{\beta}\rangle + \beta_-\delta_-) - \delta_+\left(\frac{92}{9} - \frac{7}{16}\pi^2\right) + \frac{1}{3}X_{AB}\delta_- - \frac{2}{3}\langle\bar{\kappa}\rangle - \frac{4}{3}\kappa_+ + \langle\varepsilon\rangle\Bigg\} + \nu^2(-4\beta_-^2),
\end{aligned} \tag{2.36}$$

[3] Let us remember in this context that the 1PN term in the a_i function, a_2, is identically zero in GR [37], while it is nonzero in ST theory [45].

$$\delta b^{\rm ST}_{3,\rm loc} = 29\bar{\gamma}_{AB} + \frac{131}{4}\bar{\gamma}^2_{AB} + \frac{47}{4}\bar{\gamma}^3_{AB} + 22\,(1+\bar{\gamma}_{AB})\,\langle\bar{\beta}\rangle - 5\,(1+\bar{\gamma}_{AB})\,\langle\delta\rangle - \langle\varepsilon\rangle$$
$$+ \nu\Big[- 93\bar{\gamma}_{AB} - 42\bar{\gamma}^2_{AB} - 3\bar{\gamma}^3_{AB} - 7\zeta + 2\,(3+2\bar{\gamma}_{AB})\,\delta_+ + 2\langle\delta\rangle + \varepsilon_+ + X_{AB}\varepsilon_-$$
$$- 2\left(\frac{57}{4} + \bar{\gamma}_{AB}\right) X_{AB}\beta_- - 2\,(4-9\bar{\gamma}_{AB})\,\langle\bar{\beta}\rangle\Big] + \nu^2\left(10\bar{\gamma}_{AB} + \bar{\gamma}^2_{AB} - 6\zeta - 18\beta_+ + 4\delta_+ + 2\varepsilon_+\right), \tag{2.37}$$

$$\delta q^{\rm ST}_{3,\rm loc} = \nu\left(-\frac{26}{3}\bar{\gamma}_{AB} - \frac{5}{2}\bar{\gamma}^2_{AB} - \frac{2}{3}\langle\bar{\beta}\rangle + \frac{2}{3}\langle\delta\rangle\right) + \nu^2\left(4\bar{\gamma}_{AB} - 2\langle\bar{\beta}\rangle\right), \tag{2.38}$$

where we combine the notations of Refs. [35, 45]. More precisely, introducing $X_{A,B} \equiv m^0_{A,B}/M$, we have

$$X_{AB} \equiv X_A - X_B, \tag{2.39}$$
$$\langle\bar{\beta}\rangle \equiv -X_{AB}\beta_- + \beta_+, \tag{2.40}$$
$$\langle\bar{\kappa}\rangle \equiv -X_{AB}\kappa_- + \kappa_+\ , \tag{2.41}$$
$$\langle\delta\rangle \equiv X_{AB}\delta_- + \delta_+, \tag{2.42}$$
$$\langle\bar{\chi}\rangle \equiv -X_{AB}\chi_- + \chi_+ \equiv -\frac{\langle\varepsilon\rangle}{4}, \tag{2.43}$$
$$\langle\varepsilon\rangle \equiv -X_{AB}\varepsilon_- + \varepsilon_+\ , \tag{2.44}$$

with the "$\pm$" subscript denoting the symmetric and anti-symmetric parts of the ST parameters, e.g. $x_\pm \equiv (x_A \pm x_B)/2$.

As expected, the functions $(A,\ B,\ \hat{Q}_e)$ do not depend on the function $\mathscr{F}$ of Sect. 2.3 since, similarly to the 2PN case of Ref. [45], it is absorbed by the canonical transformation. As a consistency check of our results, we verify that the binding energy of the system along circular orbits, obtain from the condition $\partial_{\hat{r}}\hat{H}_{\rm eff} = 0$, exactly matches the corresponding function given in Eq. (5.4) of Ref. [32]. Similarly, we correctly obtained the GR terms up to 3PN order as well as the ST ones at 2PN order calculated in Refs. [44, 45].

2.5.3 Completing the Circular Conservative 3PN Dynamics with Contributions from Tails and Tides

Throughout this Chapter, we have focused on the instantaneous contributions to the EOB Hamiltonian at 3PN order. However, as mentioned previously, at 3PN order two more contributions enter the PN computations: (i) the (nonlocal-in-time) tail terms [31, 32]; and (ii) the finite-size (tidal) effects [33]. The coordinate invariant circular case real two-body energy for the tail and finite-size contributions at 3PN order in ST theory are given in Eq. (5.5) of Ref. [32], and Eq. (8) of Ref. [33], respectively. In this section, we will restrict ourselves to circular orbits and compute the corrections to the EOB metric potential A due to both tail and finite-size effects. For the circular case, the *complete* 3PN term $\delta_4^{\rm ST}$ can be decomposed as

$$\delta a_4^{\rm ST} = \delta a_{4,\rm loc}^{\rm ST} + \delta a_{4,\rm nonloc}^{\rm ST} + \delta a_{4,\rm tidal}^{\rm ST} \,. \tag{2.45}$$

When only considering circular systems, the EOB metric potential A is simply computed by comparing the circular case real two-body ordinary Hamiltonian with the circular case EOB Hamiltonian using Eq. (2.24).

Let us now compute the *complete* 3PN metric potential A. To do so, we first compute the gauge invariant circular case EOB energy by considering the circular case conditions $\hat{p}_r = \frac{d\hat{p}_r}{dt} = 0$ (here, we have suppressed the subscript eff). Using these conditions in the Hamilton's equation (1.24) gives angular momentum $\hat{p}_\phi$ as function of radial distance $\hat{r}$, which can then be inverted to obtain $\hat{r}$ in terms of $\hat{p}_\phi$. This is then used in Eqs. (2.25) and (2.24) to obtain real two-body Hamiltonian and hence, the orbital angular frequency Ω using the Hamilton's equation (1.16) in terms of $\hat{p}_\phi$ and metric potential A. Following this, we use the relation $\Omega^{2/3} = x$ between the angular frequency and the gauge invariant variable x to first derive $\hat{p}_\phi$ in terms of x and metric potential A and then use it to express the real two-body Hamiltonian as a function of x and metric potential A. This gauge invariant EOB energy is then matched with the gauge invariant real two body energy using Eq. (2.24) to find ST corrections to A.

We compute the complete tail contributions of the B and Q_e corrections for noncircular orbits in Chap. 3 which is based on work published in Ref. [64].

2.5.3.1 Tail Effects

Following the procedure discussed above, the tail contribution to A reads

$$\delta a_{4,\rm nonloc}^{\rm ST} = \delta a_{4,\rm nonloc,0}^{\rm ST} + \delta a_{4,\rm nonloc,log}^{\rm ST} \log\left(\frac{1}{\hat{r}}\right), \tag{2.46}$$

where

$$\delta a_{4,\rm nonloc,0}^{\rm ST} = \nu\left(2\delta_+ + \frac{\bar{\gamma}_{AB}(2+\bar{\gamma}_{AB})}{2}\right)\left[\frac{8}{3}(\log 2 + \gamma_E)\right], \tag{2.47}$$

and

$$\delta a_{4,\rm nonloc,log}^{\rm ST} = \frac{4}{3}\nu\left(2\delta_+ + \frac{\bar{\gamma}_{AB}(2+\bar{\gamma}_{AB})}{2}\right), \tag{2.48}$$

with δ_+ defined as a symmetric part of ST parameter $\delta_{A,B}$ like above. For the equal-mass case, the tail contribution, Eq. (2.46), vanishes as the common factor of Eqs. (2.47)–(2.48) for the equal mass case is zero. This can be seen by using the definitions from Table 2.1 and that scalar charge for both bodies is same for equal mass binaries, i.e. $\alpha_A^0 = \alpha_B^0$, in the common factor of Eqs. (2.47)–(2.48).

2.5.3.2 Finite-Size Effects

When considering extended bodies, tidal effects have an impact on the binary dynamics because the tidal force exerted by the first body distorts the second body inducing in it a multipole moment excitation. The tidal effects are modelled by dimensionless parameters called tidal Love numbers which in GR are defined from the ratio between the induced quadrupole moment and the external tidal field. Now, due to the presence of a scalar field in ST theories, the tidal force exerted by the first body induces an excitation in the scalar dipolar moment of the second object. Similar to tidal Love numbers in GR, a dimensionless quantity called scalar tidal Love number is defined in ST theories to relate the excitation of the scalar dipolar moment to the external scalar tidal field [33, 65]. These tidal Love numbers are absent in GR due to the vanishing dipolar moment in COM frame. Since the first scalar tidal Love number in ST theories is due to the dipolar moment, the tidal effects appear within the conserved quantities at 3PN order instead at 5PN order in GR where the first non-vanishing tidal love number is due to the quadrupolar moment [66].

By considering the finite-size addition to the 3PN circular case energy, Eq. (8) of Ref. [33], the procedure discussed above yields

$$\delta a_{4,\text{tidal}}^{\text{ST}} = -\frac{16}{5}\nu\alpha_0^2\left[\frac{m_A^0}{M}\delta_A k_A^{(s)} + \frac{m_B^0}{M}\delta_B k_B^{(s)}\right], \tag{2.49}$$

where $k_{A,B}^{(s)}$ are the dimensionless scalar $\ell = 1$ tidal Love numbers of the two bodies. Note that BHs are expected to have a vanishing scalar Love number [65].

Now, to estimate the magnitude of scalar-mediated finite size effects, we observe that the pre-factor in Eq. (2.49) for the equal-mass case is $\mathcal{O}(10^{-8})$ in the spontaneous scalarization regime, i.e. for $\alpha \simeq 10^{-1}$. The correction to A is then linear in the scalar Love numbers, which for neutron stars typically range up to $\lesssim 10^{-2}$ [67] for $\ell = 2$ mode, depending on the equation of state. Assuming similar order of magnitude for $\ell = 1$ mode, we obtain a correction of at most $\mathcal{O}(10^{-10})$ to the metric potential A, thus rendering the effect of scalar tides on the ISCO frequency practically unmeasurable even in the spontaneous scalarization regime. However, it might be important to compute $\ell = 1$ scalar tidal love numbers for neutron stars to study the impact of finite-size effects in ST theories.

It should be noted here that in ST theories in addition to the scalar tidal Love numbers, there are also the usual tidal love numbers associated to the mass and current multipole moments as in GR. As mentioned previously, the leading order contribution to the usual multipole moments is a quadrupole moment, the tidal effects due to these start contributing at 5PN order only. Therefore, for a complete description of BNS systems in ST theories one must also take into account the quadrupolar tidal effects starting at 5PN order.

2.6 Modifications of the Conservative Binary Dynamics at the Innermost Stable Circular Orbit

Let us now use the results computed in previous sections to study the impact of the 3PN ST corrections on the circular dynamics. We do so by evaluating the frequency at the ISCO.

For the GR part, we rely on the numerical relativity (NR) informed A potential used within the TEOBResumS waveform model [68–70]. More precisely, A^{GR} is based on (formal) 5PN information where the non-logarithmic 5PN coefficient, $a_6^c(\nu)$, is informed by NR simulation *after* the full potential has been resummed using a $(1, 5)$ Padé approximant, and is given by Eq. (33) of Ref. [68]. In practice, we have

$$A_{\text{GR}}(\hat{r};\nu) \equiv P_5^1\left[A_{\text{GR}}^{5\text{PN}}(\hat{r};\nu)\right] \tag{2.50}$$

where $A_{\text{GR}}^{5\text{PN}}$ indicates the GR A function expanded at 5PN order. Focusing on the sequence of circular orbits ($\hat{p}_r = 0$), the ISCO orbital radius and angular momentum ($\hat{r}^{\text{ISCO}}$, p_ϕ^{ISCO}) are defined by the conditions $\partial_{\hat{r}}\hat{H}_{\text{eff}} = \partial_{\hat{r}}^2\hat{H}_{\text{eff}} = 0$, and Ω_{ISCO} is obtained from the corresponding Hamilton's equation. Ref. [45] already considered two ways of flexing the GR potential to include the scalar-tensor contribution. First, δA^{ST} was just considered as a contribution simply added to the Padé resummed GR potential of Eq. (2.50), but this choice was not found to be robust versus variations in the ST coupling constant. As an alternative, Ref. [45] proposed to resum, with a Padé approximant, the full PN-expanded function with GR and ST contributions. We follow this approach here and define

$$A(\hat{r}) = P_5^1\left[A_{5\text{PN}}^{\text{GR}} + \delta A^{\text{ST}}\right], \tag{2.51}$$

where now δA^{ST} includes up to the 3PN instantaneous correction and tail correction from our results of Refs. [1, 64]. Focusing on the equal-mass case, Ref. [45] noted that the 1PN and 2PN ST corrections are numerically of the same order, so that, for simplicity, in the numerical analysis of the ISCO frequency behaviour they were considered to be exactly the same. For completeness, we stick here to using the correct analytical expression without any approximation.

Figure 2.1 focuses on an equal-mass binary[4] and depicts Ω_{ISCO} as a function of the ST coupling constant α considered in a reasonable range of values compatible

[4] The ST tail correction vanishes for the equal-mass case as can be seen from the expression of the tail contribution.

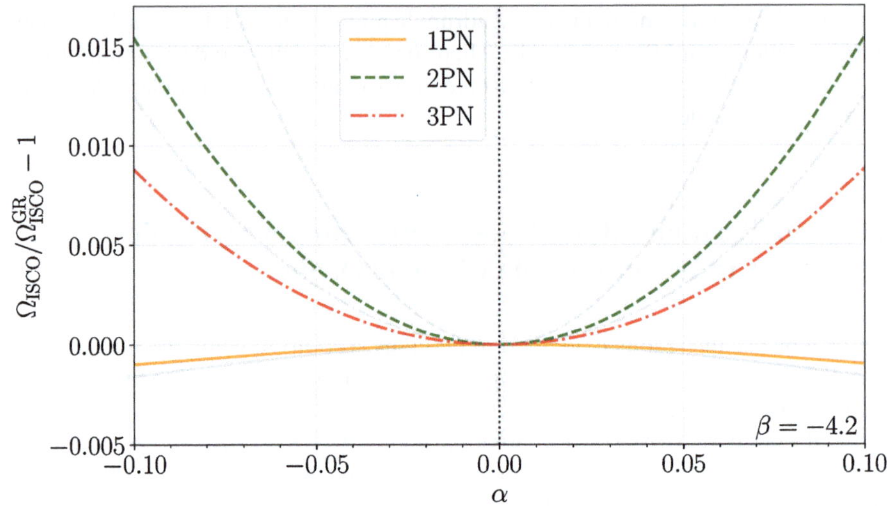

Fig. 2.1 Fractional change in the ISCO frequency with respect to GR for equal-mass systems, $\nu = 1/4$, as a function of the ST parameter α, with β fixed to $\beta = -4.2$. We show different PN orders for the ST corrections [included in the overall Padé resummation, Eq. (2.51)]. They all reduce to the GR limit when $\alpha \to 0$. The results obtained by simply adding the non-resummed ST terms are reported using grey lines

with the experimental constraints, where we fix $\beta = -4.2$. Here, for simplicity, we neglect the corrections from the ST parameters β' and β'' by fixing them to zero. The impact of varying β' and β'' within a reasonable range of values $[-10, 10]$ on purely the ST corrections to the metric potential A, $\delta a_3^{\rm ST}$ and $\delta a_4^{\rm ST}$, at the 2PN and 3PN order is 5% and 1% respectively. Therefore, their overall contribution to the potential A (see Eqs. (2.27) and (2.33)), that controls the circular conservative dynamics and thus the correction to the ISCO frequency, will indeed be negligible. From Fig. 2.1, we observe that $\Omega_{\rm ISCO}$ expectedly reduces to its GR value in the GR limit, i.e. when $\alpha \to 0$. We also observe that the ST corrections push the ISCO frequency to higher frequencies compared to GR. This implies that the binaries inspiral faster in ST theory leading to a dephasing between GWs in GR and ST theory. This dephasing effect which can be measured and hence, can be used to constrain ST parameters using GW observations.

For comparison, the figure also shows as dashed lines the curves obtained when not applying a resummation on the ST part of the potential. In Fig. 2.2 we also show the modification of $\Omega_{\rm ISCO}$ as a function of both α and β within a viable range, again for the equal-mass case.

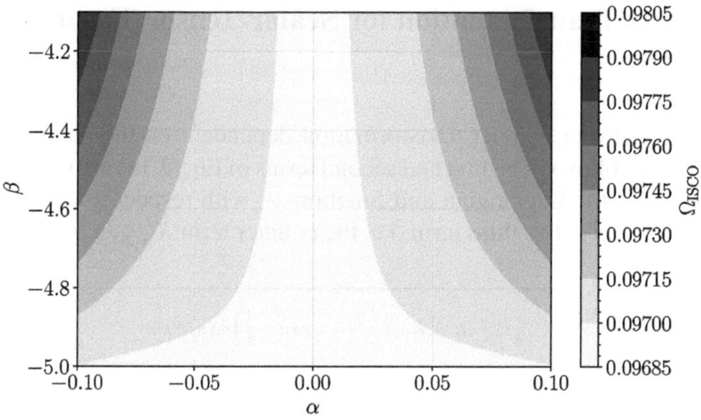

Fig. 2.2 The frequency of the innermost stable circular orbit at 3PN order for equal-mass systems, as a function of the ST parameters α and β

2.7 Conclusions

Building upon recent results of Ref. [32] for massless scalar-tensor theory, we have generalized the 2PN EOB Hamiltonian of Ref. [45] to 3PN order, though restricting ourselves to the local-in-time contribution. First, we derived a class of two-body ordinary Hamiltonians (i.e. dependent only on position and momenta) for ST theory by: (i) reducing the two-body harmonic coordinate local-in-time Lagrangian of Ref. [32] to an ordinary class of Lagrangians by constructing a general contact transformation for ST theories, (ii) augmenting the general contact transformation with the logarithm-dependent terms to gauge away regularization constants from the ordinary Lagrangian in analogy with General Relativity results of Ref. [49] and (iii) performing a Legendre transformation of this ordinary Lagrangian to derive the ordinary Hamiltonian. We then recast this local-in-time ordinary Hamiltonian into equivalent, 3PN-accurate, EOB potentials (A, B, Q_e), see Eqs. (2.36)-(2.38). As a test of our results, we checked that the energy along circular orbits computed from the EOB Hamiltonian precisely coincides with the one given in Eq. (5.4) of Ref. [32]. We also computed the corrections due to the tail and finite-size effects at 3PN for ST theories only for the circular conservative dynamics using the gauge-invariant circular case energy of Refs. [32, 33]. We additionally studied the shift in the orbital frequency of the innermost stable circular orbit induced by scalar-tensor corrections.

These results must be seen as a first step towards incorporating massless scalar-tensor corrections within the EOB waveform model `TEOBResumS-GIOTTO` [71, 72] for coalescing, precessing, BH-NS and NS-NS binaries. As mentioned previously, the nonlocal in-time corrections in the scalar-tensor theories start at 3PN order. The next step will be to derive its complete contribution at 3PN order. We will present our results of the nonlocal in-time contributions in the next Chapter. By combining the results of this Chapter and the next, we will have the corrections due to scalar-tensor theories up to 3PN order for the conservative part of the dynamics.

2.8 Contact Transformation for Scalar-Tensor Theory at 3PN

The final result of our contact transformation dependent on the parameters of the function $\mathscr{F}$ is as follows. The first and second terms of Eq. (2.14) are readily obtained by differentiating the Lagrangian and function $\mathscr{F}$ with respect to acceleration and velocity, respectively. The third term, i.e. the counter term $X'^{i}_{A,B}$, a purely 3PN order term is given by

$$
\begin{aligned}
X^i_A =\, & \vec{V}^i_B \Big\{ G_{AB} m^0_A m^0_B (\vec{V}_A\cdot\vec{V}_A)\Big[-\frac{1}{2} f_2 (\vec{n}_{AB}\cdot\vec{V}_A) + \Big(\frac{1}{2} f_5 + \frac{1}{2}\bar{\gamma}_{AB} + \frac{7}{8}\Big)\vec{n}_{AB}\cdot\vec{V}_B\Big] \\
& + \frac{G^2_{AB}(m^0_A)^2 m^0_B}{R}\Big[f_3\,(7+4\bar{\gamma}_{AB})\,\vec{n}_{AB}\cdot\vec{V}_A - f_6\,(7+4\bar{\gamma}_{AB})\,\vec{n}_{AB}\cdot\vec{V}_B\Big] \\
& + \frac{G^2_{AB} m^0_A (m^0_B)^2}{R}\Big[f_2(-3-2\bar{\gamma}_{AB})\vec{n}_{AB}\cdot\vec{V}_A + \Big(2\bar{\gamma}^2_{AB} + \frac{13\bar{\gamma}_{AB}}{2} + 2\bar{\gamma}_{AB} f_5 + 3 f_5 + \frac{21}{4}\Big)\vec{n}_{AB}\cdot\vec{V}_B\Big]\Big\} \\
& + \vec{V}^i_A \Big\{ G_{AB} m^0_A m^0_B \Big[\Big(-f_3 - \frac{1}{2}\bar{\gamma}_{AB} - \frac{7}{8}\Big)\vec{n}_{AB}\cdot\vec{V}_A(\vec{V}_B\cdot\vec{V}_B) + \Big(3 f_4 \vec{n}_{AB}\cdot\vec{V}_B - 4 f_1 \vec{n}_{AB}\cdot\vec{V}_A\Big)\vec{V}_A\cdot\vec{V}_A \\
& + \Big(-2 f_2\, \vec{n}_{AB}\cdot\vec{V}_A + \Big(f_5 + \bar{\gamma}_{AB} + \frac{7}{4}\Big)\vec{n}_{AB}\cdot\vec{V}_B\Big)\vec{V}_A\cdot\vec{V}_B - 3 f_7 (\vec{n}_{AB}\cdot\vec{V}_A)^3 - 2 f_8 (\vec{n}_{AB}\cdot\vec{V}_A)^2 \vec{n}_{AB}\cdot\vec{V}_B \\
& + (f_9 + \frac{1}{8})\vec{n}_{AB}\cdot\vec{V}_A(\vec{n}_{AB}\cdot\vec{V}_B)^2\Big] + \frac{G^2_{AB}(m^0_A)^2 m^0_B}{R}\Big[\Big(2\bar{\gamma}_{AB} f_2 + \frac{7}{2} f_2 - f_{11} + 2\bar{\gamma}^2_{AB} + 7\bar{\gamma}_{AB} + \frac{49}{8}\Big)\vec{n}_{AB}\cdot\vec{V}_A \\
& + f_5(-2\bar{\gamma}_{AB} - \frac{7}{2})\vec{n}_{AB}\cdot\vec{V}_B\Big] + \frac{G^2_{AB} m^0_A (m^0_B)^2}{R}\Big[(-f_{12} - 6 f_1 - 4\bar{\gamma}_{AB} f_1)\vec{n}_{AB}\cdot\vec{V}_A + f_4(6+4\bar{\gamma}_{AB})\vec{n}_{AB}\cdot\vec{V}_B\Big]\Big\} \\
& + \vec{n}^i_{AB}\Big\{ G_{AB} m^0_A m^0_B \Big[\Big(-\frac{3}{2} f_7 (\vec{n}_{AB}\cdot\vec{V}_A)^2 - f_8 (n_{AB}\cdot\vec{V}_A)(\vec{n}_{AB}\cdot\vec{V}_B) + \Big(\frac{f_9}{2} + \frac{1}{16}\Big)(\vec{n}_{AB}\cdot\vec{V}_B)^2 \\
& + \Big(-\frac{\bar{\gamma}_{AB}}{4} - \frac{f_3}{2} - \frac{7}{16}\Big)\vec{V}_B\cdot\vec{V}_B - \frac{1}{2} f_1(\vec{V}_A\cdot\vec{V}_A) - \frac{1}{2} f_2(\vec{V}_A\cdot\vec{V}_B)\Big)\vec{V}_A\cdot\vec{V}_A\Big] \\
& + \frac{G^2_{AB}(m^0_A)^2 m^0_B}{R}\Big[\Big(-\frac{7}{2} - \bar{\gamma}^2_{AB} - \frac{15\bar{\gamma}_{AB}}{4} - 2\bar{\gamma}_{AB} f_4 - 4 f_4 - \frac{f_{11}}{2}\Big)\vec{V}_A\cdot\vec{V}_A + f_5(-4-2\bar{\gamma}_{AB})\vec{V}_A\cdot\vec{V}_B \\
& + f_6(-4-2\bar{\gamma}_{AB})\vec{V}_B\cdot\vec{V}_B + \Big(\frac{11}{8} + \frac{3\bar{\gamma}_{AB}}{2} + \frac{1}{2} f_2 + f_8(4+2\bar{\gamma}_{AB})\Big)(\vec{n}_{AB}\cdot\vec{V}_A)^2 \\
& + \Big(f_3 - \frac{1}{2} f_5 - 4 f_9(2+\bar{\gamma}_{AB})\Big)(\vec{n}_{AB}\cdot\vec{V}_A)(\vec{n}_{AB}\cdot\vec{V}_B) + (-f_6 - 6 f_{10}(2+\bar{\gamma}_{AB}))\,(\vec{n}_{AB}\cdot\vec{V}_B)^2\Big] \\
& + \frac{G^2_{AB} m^0_A (m^0_B)^2}{R}\Big[\Big(f_1(-3-2\bar{\gamma}_{AB}) - \frac{1}{2} f_{12}\Big)\vec{V}_A\cdot\vec{V}_A + f_2(-3-2\bar{\gamma}_{AB})\vec{V}_A\cdot\vec{V}_B \\
& + \Big(-\frac{21}{8} - \frac{13\bar{\gamma}_{AB}}{4} - \bar{\gamma}^2_{AB} + f_3(-3-2\bar{\gamma}_{AB})\Big)\vec{V}_B\cdot\vec{V}_B + 3 f_7(-3-2\bar{\gamma}_{AB})(\vec{n}_{AB}\cdot\vec{V}_A)^2 \\
& + 2 f_8(-3-2\bar{\gamma}_{AB})(\vec{n}_{AB}\cdot\vec{V}_A)(\vec{n}_{AB}\cdot\vec{V}_B) + \Big(\frac{3}{8} + \frac{\bar{\gamma}_{AB}}{4} + f_9(3+2\bar{\gamma}_{AB})\Big)(\vec{n}_{AB}\cdot\vec{V}_B)^2\Big] \\
& + \frac{G^3_{AB}(m^0_A)^3 m^0_B}{R^2}\big[f_{13}(-4-2\bar{\gamma}_{AB})\big] + \frac{G^3_{AB}(m^0_A)^2(m^0_B)^2}{R^2}\big[f_{11}(-3-2\bar{\gamma}_{AB}) + f_{14}(-4-2\bar{\gamma}_{AB})\big] \\
& + \frac{G^3_{AB} m^0_A (m^0_B)^3}{R^2}\big[f_{12}(-3-2\bar{\gamma}_{AB})\big]\Big\}, \qquad (2.52)
\end{aligned}
$$

and

$$
\begin{aligned}
X_B^i = \vec{V}_B^i \Big\{ & G_{AB} m_A^0 m_B^0 \Big[\Big(f_4 + \frac{1}{2}\bar{\gamma}_{AB} + \frac{7}{8} \Big) \vec{n}_{AB} \cdot \vec{V}_B (\vec{V}_A \cdot \vec{V}_A) + \Big(4 f_6 \vec{n}_{AB} \cdot \vec{V}_B - 3 f_3 \vec{n}_{AB} \cdot \vec{V}_A \Big) \vec{V}_B \cdot \vec{V}_B \\
& + 3 f_{10} \Big(\vec{n}_{AB} \cdot \vec{V}_B \Big)^3 + \Big(2 f_5 \vec{n}_{AB} \cdot \vec{V}_B - \Big(f_2 + \bar{\gamma}_{AB} + \frac{7}{4} \Big) \vec{n}_{AB} \cdot \vec{V}_A \Big) \vec{V}_A \cdot \vec{V}_B + 2 f_9 (\vec{n}_{AB} \cdot \vec{V}_A) \Big(\vec{n}_{AB} \cdot \vec{V}_B \Big)^2 \\
& - \Big(f_8 + \frac{1}{8} \Big) \Big(\vec{n}_{AB} \cdot \vec{V}_A \Big)^2 \vec{n}_{AB} \cdot \vec{V}_B \Big] \\
& + \frac{G_{AB}^2 (m_A^0)^2 m_B^0}{R} \Big[(-4\bar{\gamma}_{AB} f_3 - 6 f_3) \vec{n}_{AB} \cdot \vec{V}_A + (4\bar{\gamma}_{AB} f_6 + 6 f_6 + f_{13}) \vec{n}_{AB} \cdot \vec{V}_B \Big] \\
& + \frac{G_{AB}^2 m_A^0 (m_B^0)^2}{R} \Big[\Big(2\bar{\gamma}_{AB} f_2 + \frac{7}{2} f_2 \Big) \vec{n}_{AB} \cdot \vec{V}_A + \Big(-2\bar{\gamma}_{AB} f_5 - \frac{7}{2} f_5 + f_{14} - 2\bar{\gamma}_{AB}^2 - 7\bar{\gamma}_{AB} - \frac{49}{8} \Big) \vec{n}_{AB} \cdot \vec{V}_B \Big] \Big\} \\
& + \vec{V}_A^i \Big\{ G_{AB} m_A^0 m_B^0 \, \vec{V}_B \cdot \vec{V}_B \Big[\Big(-\frac{1}{2} f_2 - \frac{1}{2}\bar{\gamma}_{AB} - \frac{7}{8} \Big) \vec{n}_{AB} \cdot \vec{V}_A + \frac{1}{2} f_5 \, \vec{n}_{AB} \cdot \vec{V}_B \Big] \\
& + \frac{G_{AB}^2 (m_A^0)^2 m_B^0}{R} \Big[\Big(-2\bar{\gamma}_{AB} f_2 - 3 f_2 - 2\bar{\gamma}_{AB}^2 - \frac{13}{2} - \frac{21}{4} \Big) \vec{n}_{AB} \cdot \vec{V}_A + (2\bar{\gamma}_{AB} f_5 + 3 f_5) \vec{n}_{AB} \cdot \vec{V}_B \Big] \\
& + \frac{G_{AB}^2 m_A^0 (m_B^0)^2}{R} \Big[(4\bar{\gamma}_{AB} f_1 + 7 f_1) \vec{n}_{AB} \cdot \vec{V}_A - (4\bar{\gamma}_{AB} f_4 + 7 f_4) \vec{n}_{AB} \cdot \vec{V}_B \Big] \Big\} \\
& + \vec{n}_{AB}^i \Big\{ G_{AB} m_A^0 m_B^0 \Big[\Big(-\frac{1}{2} f_8 \Big(\vec{n}_{AB} \cdot \vec{V}_A \Big)^2 + f_9 (\vec{n}_{AB} \cdot \vec{V}_B)(\vec{n}_{AB} \cdot \vec{V}_A) + \frac{3}{2} f_{10} \Big(\vec{n}_{AB} \cdot \vec{V}_B \Big)^2 + \frac{1}{2} f_4 \vec{V}_A \cdot \vec{V}_A \\
& + \frac{1}{2} f_5 \vec{V}_A \cdot \vec{V}_B + \frac{1}{2} f_6 \Big(\vec{V}_B \cdot \vec{V}_B \Big) - \frac{1}{16} \Big(\vec{n}_{AB} \cdot \vec{V}_A \Big)^2 + \frac{1}{4} \bar{\gamma}_{AB} \vec{V}_A \cdot \vec{V}_A + \frac{7}{16} \vec{V}_A \cdot \vec{V}_A \Big) \vec{V}_B \cdot \vec{V}_B \Big] \\
& + \frac{G_{AB} (m_A^0)^2 m_B^0}{R} \Big[\Big(\bar{\gamma}_{AB}^2 + \frac{13\bar{\gamma}_{AB}}{4} + (2\bar{\gamma}_{AB} + 3) f_4 + \frac{21}{8} \Big) \vec{V}_A \cdot \vec{V}_A + \Big((2\bar{\gamma}_{AB} + 3) f_6 + \frac{f_{13}}{2} \Big) \vec{V}_B \cdot \vec{V}_B \\
& + f_5 (2\bar{\gamma}_{AB} + 3) \vec{V}_A \cdot \vec{V}_B + \Big((-2\bar{\gamma}_{AB} - 3) f_8 - \frac{3}{8} - \frac{1}{4}\bar{\gamma}_{AB} \Big) \Big(\vec{n}_{AB} \cdot \vec{V}_A \Big)^2 + f_9 (4\bar{\gamma}_{AB} + 6)(\vec{n}_{AB} \cdot \vec{V}_B)(\vec{n}_{AB} \cdot \vec{V}_A) \\
& + f_{10} (6\bar{\gamma}_{AB} + 9) \Big(\vec{n}_{AB} \cdot \vec{V}_B \Big)^2 \Big] + \frac{G_{AB} m_A^0 (m_B^0)^2}{R} \Big[f_1 (2\bar{\gamma}_{AB} + 4) \vec{V}_A \cdot \vec{V}_A + f_2 (2\bar{\gamma}_{AB} + 4) \vec{V}_A \cdot \vec{V}_B \\
& + \Big(\bar{\gamma}_{AB}^2 + \frac{15\bar{\gamma}_{AB}}{4} + (2\bar{\gamma}_{AB} + 4) f_3 + \frac{f_{14}}{2} + \frac{7}{2} \Big) \vec{V}_B \cdot \vec{V}_B + ((6\bar{\gamma}_{AB} + 12) f_7 + f_1) \Big(\vec{n}_{AB} \cdot \vec{V}_A \Big)^2 \\
& + \Big((4\bar{\gamma}_{AB} + 8) f_8 + \frac{f_2}{2} - f_4 \Big) (\vec{n}_{AB} \cdot \vec{V}_A)(n_{12} \cdot \vec{V}_B) + \Big(-\frac{3\bar{\gamma}_{AB}}{4} - (2\bar{\gamma}_{AB} + 4) f_9 - \frac{f_5}{2} - \frac{11}{8} \Big) \Big(\vec{n}_{AB} \cdot \vec{V}_B \Big)^2 \Big] \\
& + \frac{G_{AB}^3 (m_A^0)^3 m_B^0}{R^2} \big[f_{13} (2\bar{\gamma}_{AB} + 3) \big] + \frac{G_{AB}^3 (m_A^0)^2 (m_B^0)^2}{R^2} \big[f_{11} (2\bar{\gamma}_{AB} + 4) + f_{14} (2\bar{\gamma}_{AB} + 3) \big] \\
& + \frac{G_{AB}^3 m_A^0 (m_B^0)^3}{R^2} \big[f_{12} (2\bar{\gamma}_{AB} + 4) \big] \Big\}.
\end{aligned}
\tag{2.53}
$$

Here, f_i are the parameters of the generic function $\mathscr{F}$ at 2PN order given in Ref. [45].

With these results, the full expression of the coordinate change (based on the contact transformation) at 3PN order, which depends on the parameters of the function $\mathscr{F}$, is derived. This is then used to obtain the class of ordinary Lagrangians in ST theory at 3PN order.

2.9 3PN Two-Body Hamiltonian

For the whole class of ordinary Hamiltonians of Sect. 2.4, the 15 coefficients h_i^{3PN} of the Hamiltonian at 3PN order for ST theory in the COM frame are

$$h_1^{3\mathrm{PN}} = -\frac{5}{128} + \frac{35}{128}\nu - \frac{70}{128}\nu^2 + \frac{35}{128}\nu^3, \qquad h_2^{3\mathrm{PN}} = h_3^{3\mathrm{PN}} = h_4^{3\mathrm{PN}} = h_5^{3\mathrm{PN}} = 0,$$

$$\begin{aligned}
h_6^{3\mathrm{PN}} = G_{AB}\Bigg\{ & -\frac{7}{16} - \frac{5}{16}\nu^3 - \frac{3}{8}\bar{\gamma}_{AB} + \frac{m_A^0}{M}(2f_6 - F_{12}) + \frac{m_B^0}{M}(2f_1 - F_1) + \nu\Bigg[4 + \frac{23}{8}\bar{\gamma}_{AB} + F_1 + F_2 - F_6 \\
& - F_7 + F_{11} + F_{12} - 2[f_1 + f_6] - \frac{3}{2}(f_2 + f_5) + \frac{3}{2}(f_3 + f_4) + \frac{m_A^0}{M}\Bigg(-F_2 + F_7 + 3F_{12} + \frac{3}{2}f_2 - \frac{3}{2}f_4 \\
& - \frac{13}{2}f_6\Bigg) + \frac{m_B^0}{M}\Bigg(3F_1 + F_6 - F_{11} - \frac{13}{2}f_1 - \frac{3}{2}f_3 + \frac{3}{2}f_5\Bigg)\Bigg] \\
& + \nu^2\Bigg[-\frac{331}{16} - \frac{93}{8}\bar{\gamma}_{AB} - 2(F_1 + F_2 - F_6 - F_7 + F_{11} + F_{12}) + \frac{9}{2}(f_1 + f_6) + \frac{7}{2}(f_2 - f_3 - f_4 + f_5) \\
& + \frac{m_A^0}{M}(F_2 + F_5 - F_7 - F_9 - F_{10} - F_{12} - 3f_2 + 3f_4 + 3f_6) + \frac{m_B^0}{M}(-F_1 - F_3 - F_4 - F_6 + F_8 + F_{11} \\
& + 3f_1 + 3f_3 - 3f_5)\Bigg]\Bigg\},
\end{aligned}$$

$$\begin{aligned}
h_7^{3\mathrm{PN}} = G_{AB}\Bigg\{ & -\frac{3}{16}\nu^3 + \frac{m_A^0}{M}[F_{12} - 3F_{24} - 2f_6 + 6f_{10}] + \frac{m_B^0}{M}[F_1 - 3F_{13} - 2f_1 + 6f_7] \\
& + \nu\Bigg[-\frac{27}{16} - \bar{\gamma}_{AB} - F_1 - F_2 + F_6 + F_7 - F_{11} - F_{12} + 3(F_{13} + F_{14} + F_{16} + F_{21} + F_{23} + F_{24}) \\
& + 2(f_1 + f_6) + \frac{3}{2}(f_2 + f_5 - f_3 - f_4) - 6(f_7 + f_{10}) - \frac{9}{2}(f_8 + f_9) \\
& + \frac{m_A^0}{M}\Bigg(F_2 - F_7 + \frac{13}{2}(f_6 - 3f_{10}) - 3\Bigg(F_{12} + F_{14} + F_{16} - 3F_{24} + \frac{1}{2}(f_2 - f_4 - 3f_8)\Bigg)\Bigg) \\
& + \frac{m_B^0}{M}\Bigg(F_{11} - F_6 - 3\Bigg(F_1 + F_{21} + F_{23} - 3F_{13} + \frac{1}{2}(f_5 - f_3 - 3f_9)\Bigg) + \frac{13}{2}(f_1 - 3f_7)\Bigg)\Bigg] \\
& + \nu^2\Bigg[\frac{169}{4} + \frac{89}{4}\bar{\gamma}_{AB} + 2(F_1 + F_2 - F_6 - F_7 + F_{11} + F_{12} - 3F_{13} - 3F_{14} - 3F_{16} - 3F_{21} - 3F_{23} - 3F_{24}) \\
& - \frac{9}{2}(f_1 + f_6 - 3f_7 - 3f_{10}) + \frac{7}{2}(f_3 - f_2 - f_5 + f_4 + 3f_8 + 3f_9) + \frac{m_A^0}{M}(-F_2 - F_5 + F_7 + F_9 + F_{10} \\
& + F_{12} + 3(F_{14} + F_{16} + F_{18} - F_{20} - F_{22} - F_{24} + f_2 - f_4 - f_6 - 3f_8 + 3f_{10})) + \frac{m_B^0}{M}(F_1 + F_3 + F_4 \\
& + F_6 - F_8 - F_{11} + 3(-F_{13} - F_{15} - F_{17} + F_{19} + F_{21} + F_{23} - f_1 - f_3 + f_5 + 3f_7 - 3f_9))\Bigg]\Bigg\},
\end{aligned}$$

$$\begin{aligned}
h_8^{3\mathrm{PN}} = G_{AB}\Bigg\{ & -\frac{3}{16}\nu^3 + \frac{m_A^0}{M}[3F_{24} - 5F_{30} - 6f_{10}] + \frac{m_B^0}{M}[3F_{13} - 5F_{25} - 6f_7] + \nu\Bigg[\frac{5}{16} + \frac{3}{8}\bar{\gamma}_{AB} \\
& - 3(F_{13} + F_{14} + F_{16} + F_{21} + F_{23} + F_{24}) + 5(F_{25} + F_{26} + F_{29} + F_{30}) + 6(f_7 + f_{10}) + \frac{9}{2}(f_8 + f_9) \\
& + 3\frac{m_A^0}{M}\Bigg(F_{14} + F_{16} - 3F_{24} - \frac{5}{3}F_{26} + 5F_{30} - \frac{3}{2}f_8 + \frac{13}{2}f_{10}\Bigg) \\
& + 3\frac{m_B^0}{M}\Bigg(F_{21} + F_{23} - 3F_{13} - \frac{5}{3}F_{29} + 5F_{25} - \frac{3}{2}f_9 + \frac{13}{2}f_7\Bigg)\Bigg] \\
& + \nu^2\Bigg[-\frac{117}{4} - \frac{129}{8}\bar{\gamma}_{AB} + 6(F_{13} + F_{14} + F_{16} + F_{21} + F_{23} + F_{24}) - 10(F_{25} + F_{26} + F_{29} + F_{30}) \\
& - \frac{21}{2}(f_8 + f_9) - \frac{27}{2}(f_7 + f_{10}) \\
& + 3\frac{m_A^0}{M}\Bigg(-F_{14} - F_{16} - F_{18} + F_{20} + F_{22} + F_{24} + \frac{5}{3}(F_{26} - F_{28} - F_{30}) + 3(f_8 - f_{10})\Bigg) \\
& + 3\frac{m_B^0}{M}\Bigg(F_{13} + F_{15} + F_{17} - F_{19} - F_{21} - F_{23} + \frac{5}{3}(F_{29} - F_{27} - F_{25}) + 3(f_9 - f_7)\Bigg)\Bigg]\Bigg\},
\end{aligned}$$

$$h_9^{3\text{PN}} = G_{AB}\left\{-\frac{5}{16}\nu^3 + 5\frac{m_A^0}{M}F_{30} + 5\frac{m_B^0}{M}F_{25} + 5\nu\left[-F_{25} - F_{26} - F_{29} - F_{30} + \frac{m_A^0}{M}(F_{26} - 3F_{30})\right.\right.$$
$$\left. + \frac{m_B^0}{M}(F_{29} - 3F_{25})\right] + 5\nu^2\left[\frac{15}{16} + \frac{1}{2}\bar{\gamma}_{AB} + 2(F_{25} + F_{26} + F_{29} + F_{30}) + \frac{m_A^0}{M}(F_{28} + F_{30} - F_{26})\right.$$
$$\left.\left. + \frac{m_B^0}{M}[F_{25} + F_{27} - F_{29}]\right]\right\},$$

$$h_{10}^{3\text{PN}} = G_{AB}^2\left\{-\frac{17\bar{\gamma}_{AB}^2}{16} - \frac{11\bar{\gamma}_{AB}}{4} - \frac{29}{16} + \frac{m_A^0}{M}[F_{12} - F_{36} + (9\bar{\gamma}_{AB} + 12)f_6 + f_{13}]\right.$$
$$+ \frac{m_B^0}{M}[F_1 - F_{41} + (9\bar{\gamma}_{AB} + 12)f_1 + f_{12}] + \nu\left[\frac{41}{12} + \frac{199}{48}\bar{\gamma}_{AB}^2 + \frac{67}{8}\bar{\gamma}_{AB} - F_1 - F_{12} + F_{36} + F_{41}\right.$$
$$+ \frac{1}{2}(f_{11} + f_{14}) - (9\bar{\gamma}_{AB} + 12)[f_1 + f_6] - f_{12} - f_{13} + \frac{m_A^0}{M}\Big(F_6 - F_{11} - 3F_{12} - F_{33} + F_{35} + 2F_{36} - F_{46}$$
$$+ 9\bar{\gamma}_{AB}(f_3 - f_5 - f_6) + \frac{1}{2}(27f_3 - 26f_5 + f_{14} - 11f_6 - 5f_{13})\Big) + \frac{m_B^0}{M}\Big(F_7 - 3F_1 - F_2 - F_{31}$$
$$\left. + F_{42} + 2F_{41} - F_{44} + 9\bar{\gamma}_{AB}(f_4 - f_1 - f_2) + \frac{1}{2}(27f_4 - 26f_2 + f_{11} - 11f_1 - 5f_{12})\Big)\right]$$
$$+ \nu^2\left[\frac{395}{48} - \frac{77\bar{\gamma}_{AB}^2}{12} - \frac{29\bar{\gamma}_{AB}}{4} + 2F_1 + F_2 - F_6 - F_7 + F_{11} + 2F_{12} + F_{31} + F_{32} + F_{33} - F_{34} - F_{35}\right.$$
$$- F_{36} - F_{41} - F_{42} - F_{43} + F_{44} + F_{45} + F_{46} + \frac{1}{2}(f_3 + f_4) - \frac{3}{2}(f_2 + f_5) - \frac{13}{2}(f_1 + f_6)$$
$$+ 2(f_{12} + f_{13} - f_{11} - f_{14}) + \frac{m_A^0}{M}\big(F_{12} - F_5 - F_6 + F_9 + F_{10} + F_{11} + 3(f_3 - f_5 - f_6)\big)$$
$$\left.\left. + \frac{m_B^0}{M}(F_1 + F_2 + F_3 + F_4 - F_7 - F_8 + 3(f_4 - f_1 - f_2))\right]\right\},$$

$$h_{11}^{3\text{PN}} = G_{AB}^2\left\{\frac{1}{16}(\bar{\gamma}_{AB} + 2)^2 + \frac{m_A^0}{M}[4F_{12} + 3F_{24} + 2F_{36} - 3F_{40} - 12f_6(2 + \bar{\gamma}_{AB}) + 9f_{10}(4 + 3\bar{\gamma}_{AB}) - 2f_{13}]\right.$$
$$+ \frac{m_B^0}{M}[4F_1 + 3F_{13} + 2F_{41} - 3F_{47} - 12f_1(2 + \bar{\gamma}_{AB}) + 9f_7(4 + 3\bar{\gamma}_{AB}) - 2f_{12}]$$
$$+ \nu\left[\frac{211}{48} + \frac{71}{8}\bar{\gamma}_{AB} + \frac{137}{48}\bar{\gamma}_{AB}^2 - 4(F_1 + F_{12}) - 3(F_{13} + F_{24} - F_{40} - F_{47}) - 2(F_{36} + F_{41}) - (f_{11} + f_{14})\right.$$
$$+ 12(2 + \bar{\gamma}_{AB})(f_1 + f_6) - 9(4 + 3\bar{\gamma}_{AB})(f_7 + f_{10}) + 2(f_{12} + f_{13}) + \frac{m_A^0}{M}\big(4(F_6 - F_{11} - 3F_{12} - F_{36})$$
$$- 3(F_{21} + F_{23} + 3F_{24} - F_{39} - 2F_{40} + F_{50}) + 2(F_{33} - F_{35} + F_{46}) - 12(2 + \bar{\gamma}_{AB})(f_3 - f_5 - f_6)$$
$$- \frac{5}{2}f_5 + \frac{19}{2}f_6 - \frac{27}{2}f_9(3 + 2\bar{\gamma}_{AB}) - \frac{27}{2}f_{10}(1 + 2\bar{\gamma}_{AB}) + 5f_{13} - f_{14}\Big) + \frac{m_B^0}{M}\big(4(F_7 - F_2 - 3F_1 - F_{41})$$
$$- 3(3F_{13} + F_{14} + F_{16} + F_{37} + 2F_{47} + 3F_{48}) + 2(F_{31} - F_{42} + F_{44}) + 12(2 + \bar{\gamma}_{AB})(f_1 + f_2 - f_4)$$
$$\left. + \frac{19}{2}f_1 - \frac{5}{2}f_2 - \frac{27}{2}f_7(1 + 2\bar{\gamma}_{AB}) - \frac{27}{2}f_8(3 + 2\bar{\gamma}_{AB}) - f_{11} + 5f_{12}\Big)\right]$$
$$+ \nu^2\left[\frac{889}{12} + 63\bar{\gamma}_{AB} + \frac{149}{12}\bar{\gamma}_{AB}^2 + 3(2F_{13} + F_{14} + F_{16} + F_{21} + F_{23} + 2F_{24} + F_{37} + F_{38} - F_{39}\right.$$
$$- F_{40} - F_{47} - F_{48} + F_{49} + F_{50}) + 4(2F_1 + F_2 - F_6 - F_7 + F_{11} + 2F_{12}) - \frac{19}{2}(f_1 + f_6) + 5(f_3 + f_4)$$
$$- \frac{45}{2}(f_7 + f_{10}) - \frac{9}{2}(f_8 + f_9) + 4(f_{11} - f_{12} - f_{13} + f_{14}) - 2(F_{31} + F_{32} + F_{33} - F_{34} - F_{35} - F_{36}$$
$$- F_{41} - F_{42} - F_{43} + F_{44} + F_{45} + F_{46}) + \frac{m_A^0}{M}\big(4(-F_5 - F_6 + F_9 + F_{10} + F_{11} + F_{12})$$

$$+\frac{9}{2}(f_3 - f_5 - f_6 - 2f_9 - 2f_{10}) + 3(-F_{18} + F_{20} + F_{21} + F_{22} + F_{23} + F_{24})\Big)$$
$$+\frac{m_B^0}{M}\Big(4(F_1 + F_2 + F_3 + F_4 - F_7 - F_8) + 3(F_{13} + F_{14} + F_{15} + F_{16} + F_{17} - F_{19})$$
$$+\frac{9}{2}(f_4 - f_1 - f_2 - 2F_7 - 2f_8)\Big)\Big]\Big\},$$

$$h_{12}^{3\text{PN}} = G_{AB}^2\Big\{\frac{m_A^0}{M}[2F_{24} + 5F_{30} + 4F_{40} - 30\bar{\gamma}_{AB}f_{10} - 48f_{10}] + \frac{m_B^0}{M}[2F_{13} + 5F_{25} + 4F_{47} - 30\bar{\gamma}_{AB}f_7 - 48f_7]$$
$$+\nu\Big[\frac{5\bar{\gamma}_{AB}^2}{36} + \frac{19\bar{\gamma}_{AB}}{12} - 2F_{13} - 2F_{24} - 5F_{25} - 5F_{30} - 4F_{40} - 4F_{47} + 30\bar{\gamma}_{AB}f_7 + 30\bar{\gamma}_{AB}f_{10} + 48f_7$$
$$+48f_{10} + \frac{227}{72} + \frac{m_B^0}{M}\Big(-6F_{13} - 2F_{14} - 2F_{16} - 15F_{25} - 5F_{26} + 4F_{37} - 8F_{47} - 4F_{48} + 30\bar{\gamma}_{AB}f_7$$
$$+30\bar{\gamma}_{AB}f_8 - f_1 + 2f_2 + \frac{81f_7}{2} + 51f_8\Big) + \frac{m_A^0}{M}\Big(-2F_{21} - 2F_{23} - 6F_{24} - 5F_{29} - 15F_{30} - 4F_{39}$$
$$-8F_{40} + 4F_{50} + 30\bar{\gamma}_{AB}f_9 + 30\bar{\gamma}_{AB}f_{10} + 2f_5 - f_6 + 51f_9 + \frac{81f_{10}}{2}\Big)\Big]$$
$$+\nu^2\Big[\frac{5\bar{\gamma}_{AB}^2}{9} - \frac{7\bar{\gamma}_{AB}}{2} + 4F_{13} + 2F_{14} + 2F_{16} + 2F_{21} + 2F_{23} + 4F_{24} + 10F_{25} + 5F_{26} + 5F_{29} + 10F_{30}$$
$$-4F_{37} - 4F_{38} + 4F_{39} + 4F_{40} + 4F_{47} + 4F_{48} - 4F_{49} - 4F_{50} - f_3 - f_4 + 6f_7 + \frac{3f_8}{2} + \frac{3f_9}{2} + 6f_{10}$$
$$-\frac{581}{144} + \frac{m_B^0}{M}\Big(2F_{13} + 2F_{14} + 2F_{15} + 2F_{16} + 2F_{17} - 2F_{19} + 5F_{25} + 5F_{26} + 5F_{27} - 2f_1 - 3f_2 - 3f_3$$
$$+f_6 + \frac{3f_9}{2} + \frac{3f_{10}}{2}\Big) + \frac{m_A^0}{M}\Big(-2F_{18} + 2F_{20} + 2F_{21} + 2F_{22} + 2F_{23} + 2F_{24} + 5F_{28} + 5F_{29} + 5F_{30}$$
$$+f_1 - 3f_4 - 3f_5 - 2f_6 + \frac{3f_7}{2} + \frac{3f_8}{2}\Big)\Big]\Big\},$$

$$h_{13}^{3\text{PN}} = \frac{G_{AB}^3\nu}{\alpha_{AB}}\Big[\frac{11}{24}(\bar{\gamma}_{AB} + 2)\bar{\gamma}_{AB} - \frac{(\bar{\gamma}_{AB} + 10)(\delta_A m_A^0 + \delta_B m_B^0)}{(6\bar{\gamma}_{AB} + 12)M}\Big] + G_{AB}^3\Big\{-\bar{\gamma}_{AB}^3 - \frac{29\bar{\gamma}_{AB}^2}{6} - \frac{47\bar{\gamma}_{AB}}{6} - \frac{17}{4}$$
$$-\frac{2(f_6 m_A^0\bar{\beta}_B + f_1 m_B^0\bar{\beta}_A)}{M} + \frac{m_A^0}{M}[F_{36} - F_{54} - 2(3\bar{\gamma}_{AB} + 5)f_6 + (5\bar{\gamma}_{AB} + 7)f_{13}]$$
$$+\frac{m_B^0}{M}[F_{41} - F_{55} - 2(3\bar{\gamma}_{AB} + 5)f_1 + (5\bar{\gamma}_{AB} + 7)f_{12}] + \frac{m_B^0\varepsilon_A + m_A^0\varepsilon_B}{12M} + [2\bar{\gamma}_{AB} + 3]\frac{m_A^0\bar{\beta}_B + m_B^0\bar{\beta}_A}{2M}$$
$$-\Big(\frac{4\bar{\gamma}_{AB}}{3} + 2\Big)\frac{m_A^0\delta_A + m_B^0\delta_B}{M}$$
$$+\nu\Big[\Big(\frac{7\pi^2}{256} - \frac{11}{24}\Big)\bar{\gamma}_{AB}^3 - \Big(\frac{5\pi^2}{64} + \frac{923}{72}\Big)\bar{\gamma}_{AB}^2 - \Big(\frac{1421}{36} + \frac{75\pi^2}{128}\Big)\bar{\gamma}_{AB} - \frac{1531}{48} - \frac{41\pi^2}{64} - F_{36} - F_{41}$$
$$+F_{54} + F_{55} + 2(3\bar{\gamma}_{AB} + 5)(f_1 + f_6) - (5\bar{\gamma}_{AB} + \frac{7}{2})(f_{12} + f_{13}) + (5\bar{\gamma}_{AB} + 7)(f_{11} + f_{14})$$
$$+\frac{m_A^0}{M}\Big(F_{33} - F_{35} - 2F_{36} + F_{46} - F_{52} - F_{53} + F_{54} + 6\bar{\gamma}_{AB}(f_5 + f_6 - f_3) - 11f_3 + \frac{21f_5}{2}$$
$$+8f_6 + \frac{3f_{11}}{2} - \frac{7f_{12}}{2}\Big) + \frac{m_B^0}{M}\Big(F_{31} - 2F_{41} - F_{42} + F_{44} - F_{51} + F_{55} - F_{56}$$
$$+6\bar{\gamma}_{AB}(f_1 + f_2 - f_4) + 8f_1 + \frac{21f_2}{2} - 11f_4 - \frac{7f_{13}}{2} + \frac{3f_{14}}{2}\Big) - \Big(7\bar{\gamma}_{AB} + \frac{37}{4}\Big)\frac{m_A^0\bar{\beta}_B + m_B^0\bar{\beta}_A}{M}$$
$$+\Big(\frac{5}{2} + \bar{\gamma}_{AB}\Big)\frac{m_A^0\bar{\beta}_A + m_B^0\bar{\beta}_B}{M} - 2f_1\frac{m_B^0\bar{\beta}_B}{M} - 2f_6\frac{m_A^0\bar{\beta}_A}{M} + 2(2f_6 + f_5 - f_3)\frac{m_A^0}{M}\bar{\beta}_B$$

$$+2(2f_1+f_2-f_4)\frac{m_B^0}{M}\bar{\beta}_A-\left(\frac{34}{9}+\frac{2}{3}\bar{\gamma}_{AB}-\frac{7}{64}\pi^2(\frac{\bar{\gamma}_{AB}}{2}+1)\right)(\delta_A+\delta_B)$$
$$+\left(\frac{\bar{\gamma}_{AB}}{6}-\frac{7}{18}\right)\frac{\delta_A m_A^0+\delta_B m_B^0}{M}+\frac{4(\bar{\beta}_A\delta_A+\bar{\beta}_B\delta_B)}{\bar{\gamma}_{AB}}-\frac{8\bar{\beta}_A\bar{\beta}_B}{\bar{\gamma}_{AB}}-\frac{m_A^0\varepsilon_A+m_B^0\varepsilon_B}{12M}\Bigg]$$
$$+\nu^2\Bigg[\frac{13\bar{\gamma}_{AB}^2}{2}+\frac{58\bar{\gamma}_{AB}}{3}+\frac{157}{12}-F_{31}-F_{32}-F_{33}+F_{34}+F_{35}+F_{36}+F_{41}+F_{42}+F_{43}-F_{44}-F_{45}$$
$$-F_{46}+f_2+f_5+2(f_1+f_6+f_{11}-f_{12}-f_{13}+f_{14})+2(\bar{\beta}_B-\bar{\beta}_A)(f_1+f_2+f_3-f_4-f_5-f_6)$$
$$-\frac{1}{2}[\bar{\beta}_A+\bar{\beta}_B]+\frac{4\bar{\beta}_A\bar{\beta}_B}{\bar{\gamma}_{AB}}+\delta_A+\delta_B+\frac{1}{12}(\varepsilon_A+\varepsilon_B)\Bigg]\Bigg\},$$

$$h_{14}^{3\mathrm{PN}}=\frac{G_{AB}^3\nu}{\alpha_{AB}}\left[-\frac{11}{4}\left(1+\frac{\bar{\gamma}_{AB}}{2}\right)\bar{\gamma}_{AB}+\frac{(\bar{\gamma}_{AB}+10)(\delta_A m_A^0+\delta_B m_B^0)}{(2\bar{\gamma}_{AB}+4)M}\right]+G_{AB}^3\Bigg\{\frac{\bar{\gamma}_{AB}^3}{4}+\frac{11\bar{\gamma}_{AB}^2}{8}+\frac{5\bar{\gamma}_{AB}}{2}+\frac{3}{2}$$
$$+\frac{m_A^0}{M}\Big[2F_{36}+3(F_{40}+F_{54})-(5+3\bar{\gamma}_{AB})(4f_6+6f_{10})-2(8+5\bar{\gamma}_{AB})f_{13}-2(2f_6+3f_{10})\bar{\beta}_B$$
$$+(\frac{3}{2}+\bar{\gamma}_{AB})\delta_A\Big]+\frac{m_B^0}{M}\Big[3(F_{47}+F_{55})-(5+3\bar{\gamma}_{AB})(4f_1+6f_7)-2(8+5\bar{\gamma}_{AB})f_{12}-2(2f_1+3f_7)\bar{\beta}_A$$
$$+2F_{41}+(\frac{3}{2}+\bar{\gamma}_{AB})\delta_B\Big]+\nu\Bigg[\left(\frac{23}{8}-\frac{21\pi^2}{256}\right)\bar{\gamma}_{AB}^3+\left(\frac{15\pi^2}{64}+\frac{713}{24}\right)\bar{\gamma}_{AB}^2+\left(\frac{225\pi^2}{128}+\frac{971}{12}\right)\bar{\gamma}_{AB}$$
$$+\frac{123\pi^2}{64}+\frac{265}{4}-2(F_{36}+F_{41})-3(F_{40}+F_{47}+F_{54}+F_{55})+(5+3\bar{\gamma}_{AB})(4(f_1+f_6)+6(f_7+f_{10}))$$
$$-2(8+5\bar{\gamma}_{AB})(f_{11}+f_{14})+\frac{5}{2}(5+4\bar{\gamma}_{AB})(f_{12}+f_{13})+\frac{m_A^0}{M}\Big(2(F_{33}-F_{35}-2F_{36}-3F_{40}+F_{46})$$
$$-3(F_{39}-F_{50}-F_{52}-F_{53}+F_{54})-(11+6\bar{\gamma}_{AB})(2f_3-3f_9)+\frac{3}{2}(15+8\bar{\gamma}_{AB})f_5+(13+12\bar{\gamma}_{AB})f_6$$
$$+3(7+6\bar{\gamma}_{AB})f_{10}-2f_{11}+\frac{7f_{12}}{2}+2(2f_1-2f_6+3f_7-3f_{10})\bar{\beta}_A+\left(8+\frac{1}{2}\bar{\gamma}_{AB}\right)\delta_A$$
$$+2\left(\frac{47}{8}+5\bar{\gamma}_{AB}-2f_3+2f_5+6f_6+3f_9+9f_{10}\right)\bar{\beta}_B+\left(\frac{77}{6}+\bar{\gamma}_{AB}\right)\delta_B-\frac{3}{4}\varepsilon_B\Big)$$
$$+\frac{m_B^0}{M}\Big(2(F_{31}-2F_{41}-F_{42}+F_{44}-3F_{47})+3(F_{37}-F_{48}+F_{51}-F_{55}+F_{56})+(13+12\bar{\gamma}_{AB})f_1$$
$$+\frac{3}{2}(15+8\bar{\gamma}_{AB})f_2+3(7+6\bar{\gamma}_{AB})f_7-(11+6\bar{\gamma}_{AB})(2f_4-3f_8)+\frac{7f_{13}}{2}-2f_{14}+\left(\frac{77}{6}+\bar{\gamma}_{AB}\right)\delta_A$$
$$+2\left(\frac{47}{8}+5\bar{\gamma}_{AB}+6f_1+2f_2-2f_4+9f_7+3f_8\right)\bar{\beta}_A+2(-2f_1+2f_6-3f_7+3f_{10})\bar{\beta}_B$$
$$+\left(8+\frac{1}{2}\bar{\gamma}_{AB}\right)\delta_B-\frac{3}{4}\varepsilon_A\Big)-\frac{21}{64}\pi^2\left(1+\frac{1}{2}\bar{\gamma}_{AB}\right)\delta_A-\frac{21}{64}\pi^2\left(1+\frac{1}{2}\bar{\gamma}_{AB}\right)\delta_B-\frac{12\bar{\beta}_A\delta_A}{\bar{\gamma}_{AB}}-\frac{12\bar{\beta}_B\delta_B}{\bar{\gamma}_{AB}}$$
$$+\frac{8\bar{\beta}_A\bar{\beta}_B}{\bar{\gamma}_{AB}}\Bigg]+\nu^2\Bigg[\frac{27\bar{\gamma}_{AB}^2}{2}+53\bar{\gamma}_{AB}+\frac{197}{4}+2(-F_{31}-F_{32}-F_{33}+F_{34}+F_{35}+F_{36}+F_{41}$$
$$+F_{42}+F_{43}-F_{44}-F_{45}-F_{46})+3(-F_{37}-F_{38}+F_{39}+F_{40}+F_{47}+F_{48}-F_{49}-F_{50})$$
$$+7f_1+2f_2-3f_3-3f_4+2f_5+7f_6+9f_7+3f_8+3f_9+9f_{10}+\frac{3}{2}(f_{12}+f_{13}-f_{11}-f_{14})$$
$$+2(\bar{\beta}_B-\bar{\beta}_A)(2f_1+2f_2+2f_3-2f_4-2f_5-2f_6+3f_7+3f_8-3f_9-3f_{10})-3(\bar{\beta}_A+\bar{\beta}_B)$$
$$+\frac{12\bar{\beta}_A\bar{\beta}_B}{\bar{\gamma}_{AB}}+3(\delta_A+\delta_B)+\frac{1}{4}(\varepsilon_A+\varepsilon_B)\Bigg]\Bigg\},$$

$$
\begin{aligned}
h_{15}^{3\mathrm{PN}} = {} & \frac{G_{AB}^4}{\alpha_{AB}}\, \nu \left[\frac{11}{12}\bar{\gamma}_{AB}(2+\bar{\gamma}_{AB}) + \left(\frac{5-\bar{\gamma}_{AB}}{6+3\bar{\gamma}_{AB}}\right)\frac{m_A^0\delta_A + m_B^0\delta_B}{M}\right] + G_{AB}^4 \left\{\frac{3}{8} + \frac{1}{3}\bar{\gamma}_{AB} + \frac{1}{12}\bar{\gamma}_{AB}^2\right. \\
& + \left(\frac{7}{6} + \frac{2}{3}\bar{\gamma}_{AB} + \frac{1}{6}\bar{\gamma}_{AB}^2\right)\frac{m_B^0\bar{\beta}_A + m_A^0\bar{\beta}_B}{M} + \frac{1}{3}\frac{m_A^0\delta_A + m_B^0\delta_B}{M} + \frac{1}{6}\frac{m_A^0\varepsilon_B + m_B^0\varepsilon_A}{M} + \frac{1}{3}\frac{m_A^0\bar{\kappa}_B + m_B^0\bar{\kappa}_A}{M} \\
& + \frac{m_A^0}{M}\left(F_{54} - 2(2+\bar{\gamma}_{AB})f_{13} + \frac{1}{2}\bar{\beta}_B^2 + \frac{2}{3}\bar{\beta}_B\delta_A - 2f_{13}\bar{\beta}_B\right) + \frac{m_B^0}{M}\left(F_{55} - 2(2+\bar{\gamma}_{AB})f_{12} + \frac{1}{2}\bar{\beta}_A^2\right. \\
& \left. + \frac{2}{3}\bar{\beta}_A\delta_B - 2f_{12}\bar{\beta}_A\right) + \nu\left[\frac{641}{21} + \frac{799}{36}\bar{\gamma}_{AB} + \frac{49}{12}\bar{\gamma}_{AB}^2 - \frac{11}{48}\bar{\gamma}_{AB}^3 + 2\bar{\beta}_A\bar{\beta}_B\left(1 - \frac{8}{\bar{\gamma}_{AB}}\right) - \frac{3}{2}\frac{m_A^0\bar{\beta}_A + m_B^0\bar{\beta}_B}{M}\right. \\
& + 2\bar{\gamma}_{AB}(-f_{11} + f_{12} + f_{13} - f_{14}) + (2f_{12} - 2f_{13})\frac{m_A^0\bar{\beta}_A - m_B^0\bar{\beta}_B}{M} - \frac{4}{3M\bar{\gamma}_{AB}}(\bar{\beta}_A\delta_A + \bar{\beta}_B\delta_B) \\
& + \left(\frac{209}{12} + \frac{20}{3}\bar{\gamma}_{AB} - \frac{1}{3}\bar{\gamma}_{AB}^2 - 2f_{11} - 2f_{14} + \frac{16\bar{\beta}_A\bar{\beta}_B}{\bar{\gamma}_{AB}^2}\right)\frac{m_A^0\bar{\beta}_B + m_B^0\bar{\beta}_A}{M} + \frac{19}{18}\frac{m_A^0\delta_B + m_B^0\delta_A}{M} \\
& + \left(\frac{5}{18} + \frac{1}{12}\bar{\gamma}_{AB}\right)\frac{m_A^0\delta_A + m_B^0\delta_B}{M} - \frac{1}{6}\frac{m_A^0\varepsilon_A + m_B^0\varepsilon_B}{M} + \frac{1}{6}\frac{m_A^0\varepsilon_B + m_B^0\varepsilon_A}{M} - \frac{1}{3}\frac{m_A^0\bar{\kappa}_A + m_B^0\bar{\kappa}_B}{M} \\
& - \frac{2}{3}\frac{m_A^0\bar{\kappa}_B + m_B^0\bar{\kappa}_A}{M} - \frac{2}{M\,\bar{\gamma}_{AB}}(m_A^0\bar{\beta}_A\varepsilon_B + m_B^0\bar{\beta}_B\varepsilon_A) + \frac{m_A^0}{M}\left(F_{52} + F_{53} - 2F_{54}\right. \\
& \left. - F_{55} - 5f_{11} + 4f_{12} + 3f_{13} - 4f_{14} + 4f_{13}\bar{\beta}_B - \frac{1}{2}\bar{\beta}_A^2 + \bar{\beta}_B^2 + \frac{2}{3}\left(\left(1 - \frac{4}{\bar{\gamma}_{AB}}\right)\bar{\beta}_A - \bar{\beta}_B\right)\delta_A - \frac{2}{3}\bar{\beta}_A\delta_B\right) \\
& + \frac{m_B^0}{M}\left(F_{51} - F_{54} - 2F_{55} + F_{56} - 4f_{11} + 3f_{12} + 4f_{13} - 5f_{14} + 4f_{12}\bar{\beta}_A + \bar{\beta}_A^2 - \frac{1}{2}\bar{\beta}_B^2 - \frac{2}{3}\bar{\beta}_B\delta_A\right. \\
& \left.\left.\left. + \frac{2}{3}\left(\left(1 - \frac{4}{\bar{\gamma}_{AB}}\right)\bar{\beta}_B - \bar{\beta}_A\right)\delta_B\right)\right]\right\}.
\end{aligned}
$$

As a consistency check with the results of GR, we find that our ST 3PN Hamiltonian reduces to the Hamiltonian of Ref. [51] in the GR limit by sctting the $\mathscr{F}_{3\mathrm{PN}}$ parameters to

$$
\begin{aligned}
& F_2 = F_{11} = 2F_{27} = 2F_{28} = 6F_{16} = 6F_{21} = -\frac{15}{8}, \\
& F_3 = F_{10} = 2F_6 = 2F_7 = \frac{2}{5}F_{34} = \frac{2}{5}F_{43} = -\frac{7}{4}, \\
& F_{33} = F_{44} = -F_{54} = -F_{55} = -\frac{2}{3}F_{15} = -\frac{2}{3}F_{22} = -\frac{3}{4}, \\
& F_5 = F_8 = \frac{47}{8}, \quad F_4 = F_9 = -\frac{23}{4}, \quad F_{14} = F_{23} = \frac{5}{12}, \\
& F_{17} = F_{20} = \frac{41}{8}, \quad F_{18} = F_{19} = -\frac{27}{16}, \quad F_{31} = F_{46} = -\frac{113}{24}, \\
& F_{32} = F_{45} = \frac{195}{16}, \quad F_{36} = F_{41} = \frac{1}{8}, \\
& F_{37} = F_{50} = -\frac{91}{144}, \quad F_{38} = F_{49} = \frac{21}{16}, \\
& F_{51} = F_{52} = \frac{245}{18} - \frac{21}{32}\pi^2, \quad F_{53} = F_{56} = -\frac{1079}{72},
\end{aligned}
$$

with the remaining parameters of $\mathscr{F}_{3\mathrm{PN}}$ set to zero and $\mathscr{F}_{2\mathrm{PN}}$ coinciding with the expression in Ref. [45].

References

1. Jain T, Rettegno P, Agathos M, Nagar A, Turco L (2023) Effective-one-body Hamiltonian in scalar-tensor gravity at third post-Newtonian order. Phys Rev D 107:084017. arXiv:2211.15580. https://doi.org/10.1103/PhysRevD.107.084017 [gr-qc]
2. Damour T, Esposito-Farese G (1992) Tensor multiscalar theories of gravitation. Class Quant Grav 9:2093–2176. https://doi.org/10.1088/0264-9381/9/9/015
3. Damour T, Esposito-Farese G (1993) Nonperturbative strong field effects in tensor - scalar theories of gravitation. Phys Rev Lett 70:2220–2223. https://doi.org/10.1103/PhysRevLett.70.2220
4. Damour T, Esposito-Farese G (1996) Testing gravity to second postNewtonian order: a Field theory approach. Phys Rev D 53:5541–5578. arXiv:gr-qc/9506063. https://doi.org/10.1103/PhysRevD.53.5541
5. Freire PCC, Wex N, Esposito-Farèse G, Verbiest JP, Bailes M, Jacoby BA, Kramer M, Stairs IH, Antoniadis J, Janssen GH (2012) The relativistic pulsar-white dwarf binary PSR J1738+0333 II. The most stringent test of scalar-tensor gravity. Mon Not Roy Astron Soc 423:3328. arXiv:1205.1450. https://doi.org/10.1111/j.1365-2966.2012.21253.x [astro-ph.GA]
6. Khalil M, Mendes RFP, Ortiz N, Steinhoff J (2022) Effective-action model for dynamical scalarization beyond the adiabatic approximation. arXiv:2206.13233 [gr-qc]
7. Gautam T, Freire PCCC, Batrakov A, Kramer M, Miao CC, Parent E, Zhu WW (2022) Relativistic effects in a mildly recycled pulsar binary: PSR J1952+2630. arXiv:2210.03464 [astro-ph.HE]
8. Doneva DD, Ramazanoğlu FM, Silva HO, Sotiriou TP, Yazadjiev SS (2024) Spontaneous scalarization. Rev Mod Phys 96:015004. arXiv:2211.01766. https://doi.org/10.1103/RevModPhys.96.015004 [gr-qc]
9. Abbott BP et al (collaboration Virgo, LIGO Scientific) (2016) Observation of gravitational waves from a binary black hole merger. Phys. Rev. Lett. 116:061102. arXiv:1602.03837. https://doi.org/10.1103/PhysRevLett.116.061102 [gr-qc]
10. Arun KG, Iyer BR, Qusailah MSS, Sathyaprakash BS (2006) Testing post-Newtonian theory with gravitational wave observations. Class Quant Grav 23:L37–L43. arXiv:gr-qc/0604018. https://doi.org/10.1088/0264-9381/23/9/L01
11. Mishra CK, Arun KG, Iyer BR, Sathyaprakash BS (2010) Parametrized tests of post-Newtonian theory using advanced LIGO and Einstein telescope. Phys Rev D 82:064010. arXiv:1005.0304. https://doi.org/10.1103/PhysRevD.82.064010 [gr-qc]
12. Li T, Del Pozzo W, Vitale S, Van Den Broeck C, Agathos M et al (2012) Towards a generic test of the strong field dynamics of general relativity using compact binary coalescence. Phys Rev D 85:082003. arXiv:1110.0530. https://doi.org/10.1103/PhysRevD.85.082003 [gr-qc]
13. Agathos M, Del Pozzo W, Li TGF, Broeck CVD, Veitch J et al (2014) TIGER: a data analysis pipeline for testing the strong-field dynamics of general relativity with gravitational wave signals from coalescing compact binaries. Phys Rev D 89:082001. arXiv:1311.0420. https://doi.org/10.1103/PhysRevD.89.082001 [gr-qc]
14. Cornish N, Sampson L, Yunes N, Pretorius F (2011) Gravitational wave tests of general relativity with the parameterized post-Einsteinian framework. Phys Rev D 84:062003. arXiv:1105.2088. https://doi.org/10.1103/PhysRevD.84.062003 [gr-qc]
15. Berti E et al (2015) Testing general relativity with present and future astrophysical observations. Class Quant Grav 32:243001. arXiv:1501.07274. https://doi.org/10.1088/0264-9381/32/24/243001 [gr-qc]
16. Yunes N, Yagi K, Pretorius F (2016) Theoretical physics implications of the binary black-hole mergers GW150914 and GW151226. Phys Rev D 94:084002. arXiv:1603.08955. https://doi.org/10.1103/PhysRevD.94.084002 [gr-qc]
17. Abbott BP et al (Collaboration LIGO Scientific, Virgo) (2016) Tests of general relativity with GW150914. Phys Rev Lett 116:221101. [Erratum: Phys. Rev. Lett.121,no.12,129902(2018)]. arXiv:1602.03841. https://doi.org/10.1103/PhysRevLett.116.221101, https://doi.org/10.1103/PhysRevLett.121.129902 [gr-qc]

18. Aasi J et al (Collaboration LIGO Scientific) (2015) Advanced LIGO. Class Quant Grav 32:074001. arXiv:1411.4547. https://doi.org/10.1088/0264-9381/32/7/074001 [gr-qc]
19. Acernese F et al (Collaboration VIRGO) (2015) Advanced Virgo: a second-generation interferometric gravitational wave detector. Class Quant Grav 32:024001. arXiv:1408.3978. https://doi.org/10.1088/0264-9381/32/2/024001 [gr-qc]
20. Akutsu T et al (Collaboration KAGRA) (2021) Overview of KAGRA: calibration, detector characterization, physical environmental monitors, and the geophysics interferometer. PTEP 2021:05A102. arXiv:2009.09305. https://doi.org/10.1093/ptep/ptab018 [gr-qc]
21. Abbott BP et al (Collaboration VIRGO, KAGRA, LIGO Scientific) (2018) Prospects for observing and localizing gravitational-wave transients with advanced LIGO, advanced Virgo and KAGRA. Living Rev Rel 21(3) [Living Rev. Rel. 19,1(2016)]. arXiv:1304.0670. https://doi.org/10.1007/s41114-018-0012-9, https://doi.org/10.1007/lrr-2016-1 [gr-qc]
22. Abbott BP et al (Collaboration LIGO Scientific, Virgo) (2019) Tests of general relativity with the binary black hole signals from the LIGO-Virgo catalog GWTC-1. Phys Rev D 100:104036. arXiv:1903.04467. https://doi.org/10.1103/PhysRevD.100.104036 [gr-qc]
23. Abbott BP et al (Collaboration LIGO Scientific, Virgo) (2019) Tests of general relativity with GW170817. Phys Rev Lett 123:011102. arXiv:1811.00364. https://doi.org/10.1103/PhysRevLett.123.011102 [gr-qc]
24. Abbott R et al (Collaboration LIGO Scientific, Virgo) (2021) Tests of general relativity with binary black holes from the second LIGO-Virgo gravitational-wave transient catalog. Phys Rev D 103:122002. arXiv:2010.14529. https://doi.org/10.1103/PhysRevD.103.122002 [gr-qc]
25. Abbott R et al (Collaboration LIGO Scientific, VIRGO, KAGRA) (2021) Tests of general relativity with GWTC-3. arXiv:2112.06861 [gr-qc]
26. Vallisneri M, Yunes N (2013) Stealth bias in gravitational-wave parameter estimation. Phys Rev D 87:102002. arXiv:1301.2627. https://doi.org/10.1103/PhysRevD.87.102002 [gr-qc]
27. Maggiore M et al (2019) Science case for the Einstein telescope. JCAP 03:050. arXiv:1912.02622. https://doi.org/10.1088/1475-7516/2020/03/050 [astro-ph.CO]
28. Evans M et al (2021) A horizon study for cosmic explorer: science, observatories, and community. arXiv:2109.09882 [astro-ph.IM]
29. Lang RN (2014) Compact binary systems in scalar-tensor gravity. II. Tensor gravitational waves to second post-Newtonian order. Phys Rev D 89:084014. arXiv:1310.3320. https://doi.org/10.1103/PhysRevD.89.084014 [gr-qc]
30. Lang RN (2015) Compact binary systems in scalar-tensor gravity. III. Scalar waves and energy flux. Phys Rev D 91:084027. arXiv:1411.3073. https://doi.org/10.1103/PhysRevD.91.084027 [gr-qc]
31. Bernard L (2018) Dynamics of compact binary systems in scalar-tensor theories: equations of motion to the third post-Newtonian order. Phys Rev D 98:044004. arXiv:1802.10201. https://doi.org/10.1103/PhysRevD.98.044004 [gr-qc]
32. Bernard L (2019) Dynamics of compact binary systems in scalar-tensor theories: II. Center-of-mass and conserved quantities to 3PN order. Phys Rev D 99:044047. arXiv:1812.04169. https://doi.org/10.1103/PhysRevD.99.044047 [gr-qc]
33. Bernard L (2020) Dipolar tidal effects in scalar-tensor theories. Phys Rev D 101:021501. arXiv:1906.10735. https://doi.org/10.1103/PhysRevD.101.021501 [gr-qc]
34. Schön O, Doneva DD (2022) Tensor-multiscalar gravity: equations of motion to 2.5 post-Newtonian order. Phys Rev D 105:064034. arXiv:2112.07388. https://doi.org/10.1103/PhysRevD.105.064034 [gr-qc]
35. Sennett N, Marsat S, Buonanno A (2016) Gravitational waveforms in scalar-tensor gravity at 2PN relative order. Phys Rev D 94:084003. arXiv:1607.01420. https://doi.org/10.1103/PhysRevD.94.084003 [gr-qc]
36. Bernard L, Blanchet L, Trestini D (2022) Gravitational waves in scalar-tensor theory to one-and-a-half post-Newtonian order. JCAP 08(08):008. arXiv:2201.10924. https://doi.org/10.1088/1475-7516/2022/08/008 [gr-qc]
37. Buonanno A, Damour T (1999) Effective one-body approach to general relativistic two-body dynamics. Phys Rev D 59:084006. arXiv:gr-qc/9811091. https://doi.org/10.1103/PhysRevD.59.084006

38. Buonanno A, Damour T (2000) Transition from inspiral to plunge in binary black hole coalescences. Phys Rev D 62:064015. arXiv:gr-qc/0001013. https://doi.org/10.1103/PhysRevD.62.064015
39. Damour T, Jaranowski P, Schaefer G (2000) On the determination of the last stable orbit for circular general relativistic binaries at the third post-Newtonian approximation. Phys Rev D 62:084011. arXiv:gr-qc/0005034. https://doi.org/10.1103/PhysRevD.62.084011 [gr-qc]
40. Damour T, Jaranowski P, Schäfer G (2008) Effective one body approach to the dynamics of two spinning black holes with next-to-leading order spin-orbit coupling. Phys Rev D 78:024009. arXiv:0803.0915. https://doi.org/10.1103/PhysRevD.78.024009 [gr-qc]
41. Damour T, Jaranowski P, Schäfer G (2014) Nonlocal-in-time action for the fourth post-Newtonian conservative dynamics of two-body systems. Phys Rev D 89:064058. arXiv:1401.4548. https://doi.org/10.1103/PhysRevD.89.064058 [gr-qc]
42. Damour T, Jaranowski P, Schäfer G (2015) Fourth post-Newtonian effective one-body dynamics. Phys Rev D 91:084024. arXiv:1502.07245. https://doi.org/10.1103/PhysRevD.91.084024 [gr-qc]
43. Damour T, Jaranowski P, Schäfer G (2016) Conservative dynamics of two-body systems at the fourth post-Newtonian approximation of general relativity. Phys Rev D 93:084014. arXiv:1601.01283. https://doi.org/10.1103/PhysRevD.93.084014 [gr-qc]
44. Julié F-L (2018) Reducing the two-body problem in scalar-tensor theories to the motion of a test particle: a scalar-tensor effective-one-body approach. Phys Rev D 97:024047. arXiv:1709.09742. https://doi.org/10.1103/PhysRevD.97.024047 [gr-qc]
45. Julié F-L, Deruelle N (2017) Two-body problem in scalar-tensor theories as a deformation of general relativity: an Effective-One-Body approach. Phys Rev D 95:124054. arXiv:1703.05360. https://doi.org/10.1103/PhysRevD.95.124054 [gr-qc]
46. Eardley DM (1975) Observable effects of a scalar gravitational field in a binary pulsar. apjl 196:L59–L62. https://doi.org/10.1086/181744
47. Schafer G (1984) Acceleration-dependent Lagrangians in general relativity. Phys Lett A 100:128–129. https://doi.org/10.1016/0375-9601(84)90947-2
48. Damour T, Schäfer G (1985) Lagrangians for n point masses at the second post-Newtonian approximation of general relativity. Gen Rel Grav 17:879–905. https://doi.org/10.1007/BF00773685
49. de Andrade VC, Blanchet L, Faye G (2001) Third post-Newtonian dynamics of compact binaries: Noetherian conserved quantities and equivalence between the harmonic coordinate and ADM Hamiltonian formalisms. Class Quant Grav 18:753–778. arXiv:gr-qc/0011063. https://doi.org/10.1088/0264-9381/18/5/301
50. Damour T, Schaefer G (1991) Redefinition of position variables and the reduction of higher order Lagrangians. J Math Phys 32:127–134. https://doi.org/10.1063/1.529135
51. Damour T, Jaranowski P, Schaefer G (2000) Dynamical invariants for general relativistic two-body systems at the third post-Newtonian approximation. Phys Rev D 62:044024. arXiv:gr-qc/9912092. https://doi.org/10.1103/PhysRevD.62.044024
52. Blanchet L, Faye G (2001) General relativistic dynamics of compact binaries at the third post-Newtonian order. Phys Rev D 63:062005. arXiv:gr-qc/0007051. https://doi.org/10.1103/PhysRevD.63.062005
53. Hinderer T et al (2013) Periastron advance in spinning black hole binaries: comparing effective-one-body and numerical relativity. Phys Rev D 88:084005. arXiv:1309.0544. https://doi.org/10.1103/PhysRevD.88.084005
54. Blümlein J, Maier A, Marquard P, Schäfer G (2022) The fifth-order post-Newtonian Hamiltonian dynamics of two-body systems from an effective field theory approach. Nucl Phys B 983:115900. arXiv:2110.13822. https://doi.org/10.1016/j.nuclphysb.2022.115900 [gr-qc]
55. Damour T (2016) Gravitational scattering, post-Minkowskian approximation and effective one-body theory. Phys Rev D 94:104015. arXiv:1609.00354. https://doi.org/10.1103/PhysRevD.94.104015 [gr-qc]
56. Damour T (2018) High-energy gravitational scattering and the general relativistic two-body problem. Phys Rev D 97:044038. arXiv:1710.10599. https://doi.org/10.1103/PhysRevD.97.044038 [gr-qc]

57. Damour T (2020) Classical and quantum scattering in post-Minkowskian gravity. Phys Rev D 102:024060. arXiv:1912.02139. https://doi.org/10.1103/PhysRevD.102.024060 [gr-qc]
58. Antonelli A, Buonanno A, Steinhoff J, van de Meent M, Vines J (2019) Energetics of two-body Hamiltonians in post-Minkowskian gravity. Phys Rev D 99:104004. arXiv:1901.07102. https://doi.org/10.1103/PhysRevD.99.104004 [gr-qc]
59. Antonelli A, van de Meent M, Buonanno A, Steinhoff J, Vines J (2020) Quasicircular inspirals and plunges from nonspinning effective-one-body Hamiltonians with gravitational self-force information. Phys Rev D 101:024024. arXiv:1907.11597. https://doi.org/10.1103/PhysRevD.101.024024 [gr-qc]
60. Khalil M, Buonanno A, Steinhoff J, Vines J (2022) Energetics and scattering of gravitational two-body systems at fourth post-Minkowskian order. Phys Rev D 106:024042. arXiv:2204.05047. https://doi.org/10.1103/PhysRevD.106.024042 [gr-qc]
61. Damour T, Rettegno P (2022) Strong-field scattering of two black holes: numerical Relativity meets post-Minkowskian gravity. arXiv:2211.01399 [gr-qc]
62. Bini D, Damour T, Geralico A (2020) Sixth post-Newtonian local-in-time dynamics of binary systems. Phys Rev D 102:024061. arXiv:2004.05407. https://doi.org/10.1103/PhysRevD.102.024061 [gr-qc]
63. Bini D, Damour T, Geralico A (2020) Sixth post-Newtonian nonlocal-in-time dynamics of binary systems. Phys Rev D 102:084047. arXiv:2007.11239. https://doi.org/10.1103/PhysRevD.102.084047 [gr-qc]
64. Jain T (2023) Nonlocal-in-time effective one body Hamiltonian in scalar-tensor gravity at third post-Newtonian order. Phys Rev D 107:084018. arXiv:2301.01070. https://doi.org/10.1103/PhysRevD.107.084018 [gr-qc]
65. Cardoso V, Franzin E, Maselli A, Pani P, Raposo G (2017) Testing strong-field gravity with tidal Love numbers. Phys Rev D 95:084014. [Addendum: Phys.Rev.D 95, 089901 (2017)]. arXiv:1701.01116. https://doi.org/10.1103/PhysRevD.95.084014 [gr-qc]
66. Damour T, Nagar A (2009) Relativistic tidal properties of neutron stars. Phys Rev D 80:084035. arXiv:0906.0096. https://doi.org/10.1103/PhysRevD.80.084035 [gr-qc]
67. Brown SM (2022) Tidal deformability of neutron stars in scalar-tensor theories of gravity for gravitational wave analysis. arXiv:2210.14025 [gr-qc]
68. Nagar A, Riemenschneider G, Pratten G, Rettegno P, Messina F (2020) Multipolar effective one body waveform model for spin-aligned black hole binaries. Phys Rev D 102:024077. arXiv:2001.09082. https://doi.org/10.1103/PhysRevD.102.024077 [gr-qc]
69. Schmidt S, Breschi M, Gamba R, Pagano G, Rettegno P, Riemenschneider G, Bernuzzi S, Nagar A, Del Pozzo W (2021) Machine learning gravitational waves from binary black hole mergers. Phys Rev D 103:043020. arXiv:2011.01958. https://doi.org/10.1103/PhysRevD.103.043020 [gr-qc]
70. Riemenschneider G, Rettegno P, Breschi M, Albertini A, Gamba R, Bernuzzi S, Nagar A (2021) Assessment of consistent next-to-quasicircular corrections and postadiabatic approximation in effective-one-body multipolar waveforms for binary black hole coalescences. Phys Rev D 104:104045. arXiv:2104.07533. https://doi.org/10.1103/PhysRevD.104.104045 [gr-qc]
71. Gamba R, Akçay S, Bernuzzi S, Williams J (2022) Effective-one-body waveforms for precessing coalescing compact binaries with post-Newtonian twist. Phys Rev D 106:024020. arXiv:2111.03675. https://doi.org/10.1103/PhysRevD.106.024020 [gr-qc]
72. Gamba R, Bernuzzi S (2023) Resonant tides in binary neutron star mergers: analytical-numerical relativity study. Phys Rev D 107:044014. arXiv:2207.13106. https://doi.org/10.1103/PhysRevD.107.044014 [gr-qc]

Chapter 3
Nonlocal-In-Time Effective-One-Body Hamiltonian in Scalar-Tensor Theories at Third Post-Newtonian Order

We derive the corrections to effective-one-body (EOB) metric potentials for the tail (nonlocal-in-time) nonlocal-in-time of conservative dynamics for massless Scalar-Tensor (ST) theories at third post-Newtonian (PN) order. The nonlocal-in- time EOB Hamiltonian is obtained by mapping the order-reduced Hamiltonian corresponding to the nonlocal-in-time Lagrangian derived. To transcribe the dynamics within EOB formalism, we use a strategy of order-reduction of nonlocal dynamics to local ordinary action-angle Hamiltonian. We then map this onto the EOB Hamiltonian to determine the nonlocal-in-time ST corrections to the EOB potentials at 3PN order.

This is based on the work published in Ref. [1].

3.1 Introduction

In this Chapter as well we derive the EOB Hamiltonian for the simplest well-posed theory, i.e., the scalar-tensor (ST) theories, which are thoroughly studied and tested [2–7]. The important violations for ST theories arise through the non-perturbative strong field effects in neutron stars such as spontaneous scalarization [4]. The most stringent constraints for ST theories come from binary pulsar observations. There is hope that future gravitational wave (GW) detections of coalescing compact binaries will complement current studies on constraints using genuine strong-field information and additional terms in the radiation, i.e. dipolar radiation which is not present in GR, due to the scalar extension of GR [6, 8, 9].

In contrast to the tail effects in GR, the tail effects in ST theory start at 3PN order [10, 11]. Therefore, to construct the *complete* Hamiltonian within the EOB formalism at 3PN order, it is important to recast the nonlocal-in-time post-Newtonian information in the EOB Hamiltonian. This Chapter should be considered as a part of a considerate effort to construct the waveform templates for ST theories by recasting the PN results within the effective-one-body (EOB) description of the two-body

T. Jain, *Surfing the Gravitational Waves*, Springer Theses,
https://doi.org/10.1007/978-3-031-96180-9_3

problem [12–18]. This Chapter aims to derive the corrections to the EOB metric potentials due to these nonlocal-in-time effects, starting from the 3PN nonlocal-in-time Lagrangian in ST of Refs. [10, 11], following the results obtained in Chap. 2 for the local dynamics in ST theory at 3PN order.

The Chapter is organised as follows. In Sect. 3.2, we briefly summarise the ST conventions and EOB notations used in this Chapter following the results of Chap. 2. Then, in Sect. 3.3 we derive the conserved energy for a nonlocal-in-time part using two methods, (i) a non-order-reduced nonlocal Hamiltonian using nonlocal phase shift, and (ii) the order-reduction of nonlocal dynamics to a local ordinary action-angle Hamiltonian. Finally, in Sect. 3.4 we map the nonlocal-in-time ordinary Hamiltonian into an EOB Hamiltonian at 3PN order.

3.2 Scalar-Tensor Parameters and Effective-One-Body Conventions

The massless ST theories considering a single scalar field are described by the following action in the Einstein Frame (i.e. when the scalar field minimally couples to the metric),

$$S = \frac{c^4}{16\pi G}\int d^4x\sqrt{-g}(R - 2g^{\mu\nu}\partial_\mu\varphi\partial_\nu\varphi) + S_m[\Psi, \mathcal{A}(\varphi)^2 g_{\mu\nu}], \tag{3.1}$$

where $g_{\mu\nu}$ is the Einstein metric, R is the Ricci scalar, φ is the scalar field, Ψ collectively denotes the matter fields, $g \equiv \det(g_{\mu\nu})$ and G is the Newtonian constant.

Following Chap. 2, here also, we work in DEF conventions of Refs. [2, 4] for the binary parameters to encompass the scalar field effect. The conversion of Jordan frame parameters used in Ref. [11] to express the nonlocal-in-time (tail) Lagrangian to the dimensionless Einstein frame parameters used in this Chapter is given Table 2.1. We also use the same notation as [10, 11] to denote weak-field and strong-field parameters.

Let us recall that at 3PN order, the EOB Hamiltonian is [14]

$$\hat{H}_{\text{eff}} = \frac{H_{\text{eff}}}{\mu} = \sqrt{A(\hat{r})\left(1 + \frac{\hat{p}_r^2}{B(\hat{r})} + \frac{\hat{p}_\phi^2}{\hat{r}^2} + \hat{Q}_e\right)}, \tag{3.2}$$

where $\hat{p}_r$, $\hat{p}_\phi$ are the dimensionless radial and angular momenta, and $\hat{r}(= r/(G_{AB}M)$ is the dimensionless radial separation and (A, B, Q_e) are the EOB metric potentials at 3PN order (hereafter, the superscript *hat* is used to denote the dimensionless variables). As mentioned in Chap. 2, there is a gauge freedom for metric potential $\hat{Q}_e$. However, similar to Chap. 2 here also, we work in DJS gauge [14]. Therefore, the three EOB potentials $\left(A(\hat{r}), B(\hat{r}), Q_e\right)$ at 3PN order formally read

$$A(\hat{r}) = 1 - \frac{2}{\hat{r}} + \frac{a_2}{\hat{r}^2} + \frac{a_3}{\hat{r}^3} + \frac{a_4}{\hat{r}^4}, \tag{3.3}$$

$$B(\hat{r}) = 1 + \frac{b_1}{\hat{r}} + \frac{b_2}{\hat{r}^2} + \frac{b_3}{\hat{r}^3}, \tag{3.4}$$

$$\hat{Q}_e(\hat{r}) = q_3 \frac{\hat{p}_r^4}{\hat{r}^2}. \tag{3.5}$$

The GR and ST corrections in terms of the symmetric mass-ratio (ν)-dependent coefficients (a_i, b_i) are then separated as

$$a_i = a_i^{\text{GR}} + \delta a_i^{\text{ST}}, \tag{3.6}$$

$$b_i = b_i^{\text{GR}} + \delta b_i^{\text{ST}}, \tag{3.7}$$

$$q_3 = q_3^{\text{GR}} + \delta q_3^{\text{ST}}. \tag{3.8}$$

Since there are local-in-time, nonlocal-in-time and tidal contributions at 3PN order in ST theory, all the 3PN ST coefficients can thus be decomposed as Eq. (2.45) of Chap. 2. In this Chapter, we will focus on computing the nonlocal-in-time corrections to these three EOB potentials at 3PN order.

3.3 Tail Contribution at the 3PN Order to the Two-Body Dynamics

The nonlocal-in-time two-body 3PN Lagrangian for massless ST theory obtained in Ref. [11] is in harmonic coordinates, i.e. it depends (linearly) on the acceleration of the two bodies. In this section, we use two different methods to derive the *Noetherian* conserved energy for the tail contributions.

In the first method, we remove the acceleration dependence from the Lagrangian (hence, the Hamiltonian) and stay within the non-order-reduced nonlocal framework as done in Refs. [19, 20] for GR. However, in this framework, one must *not* consider Hamiltonian as conserved energy. The difference between the conserved energy and Hamiltonian within non-local framework can be understood by considering a toy model in which non-local effects decay sufficiently fast such that the non-local Hamiltonian can be expressed as a Taylor series. The Taylor series expansion of the non-local Hamiltonian would involve higher derivatives of phase-space variables implying that it is replaced by a generalized Ostrogradski Hamiltonian (i.e., the equations of motion in Lagrangian Formalism are dependent on more than two time-derivatives). In Refs. [18, 21], it has been shown that for an Ostrogradski Hamiltonian, the conserved energy is not equal to the Hamiltonian but is a quantity derived from the Hamiltonian.

In the second method, we derive the order-reduced, local Hamiltonian using the action-angle variables (see, Ref. [17] for GR). Contrary to the first method of non-local framework, where the conserved energy is *not* equal to the Hamiltonian, the order-reduced local Hamiltonian obtained using action-angle variables is equal to the conserved energy. This self-consistency of the order-reduced local Hamiltonian is due to the fact that we can use the "on-shell" equations of motion, i.e., the non-local shifts entering as source in action-angle variables can be "redefined-away" [18]. Due to this consistency between the conserved energy and the Hamiltonian, the second method is preferable in contrast to the first method. Additionally, the second method is more useful in simplifying the mapping between the two-body dynamics and the EOB dynamics which is the main aim of this chapter. This is because the non-local dynamics cannot be mapped to the EOB dynamics by the usual canonical transformation considered in Chap. 2. There could be two ways to incorporate the non-locality in the EOB formalism: (i) to introduce a non-local structure in the EOB Hamiltonian by introducing some non-locality in the effective metric $g_{\mu\nu}^{\rm eff}$ and the non-geodesic term Q or (ii) to replace the non-locality by local contributions in the effective metric $g_{\mu\nu}^{\rm eff}$ and infinitely many local contributions to the non-geodesic term Q. As argued in Ref. [17], the second approach of incorporating the non-locality into the EOB dynamics is more useful which requires the order-reduction of the non-local Hamiltonian into local Hamiltonian as explained in the second method of constructing conserved energy.

3.3.1 Non-order-Reduced Ordinary Hamiltonian

In Chap. 2, we have derived the ordinary (dependent only on positions and momenta) Hamiltonian for local-in-time contributions using a contact transformation (see, Appendix 2.8 of Chap. 2 for the contact transformation). Now, due to the presence of a nonlocal piece in the nonlocal-in-time part of the Lagrangian, we need to find the nonlocal coordinate shift that removes the acceleration dependence from the tail part of the Lagrangian of Ref. [11] (see, Refs. [19, 20] for GR). Corresponding to this ordinary Lagrangian, we can then derive the ordinary Hamiltonian.

Therefore, the total Lagrangian at 3PN order in ST can be written as [11],

$$L = L^{\rm inst} + L^{\rm tail}, \tag{3.9}$$

where $L^{\rm inst}$ and $L^{\rm tail}$ are the instantaneous and tail contributions to the Lagrangian, respectively. As mentioned previously, the instantaneous contribution is already considered in Chap. 2 and here we focus on the tail contributions only. The tail part of the Lagrangian at 3PN order reads,

$$L^{\rm tail} = \frac{2G^2M}{3c^6}(3+2\omega_0)\,{\rm Pf}_{2r_{AB}/c}\int_{-\infty}^{\infty}\frac{d\tau}{|\tau|}I_{s,i}^{(2)}(t)I_{s,i}^{(2)}(t+\tau), \tag{3.10}$$

where Pf is the Hadamard partie finie function, the Hadamard scale $r_{AB}(=r)$ is the relative separation of two bodies, and $I_{s,i}^{(2)}$ is the second time derivative of the scalar dipole moment, $I_{s,i}$. Here, we find the coordinate shift that transforms this Lagrangian into the same expression but with the derivatives of the dipole moment evaluated using the Newtonian equations of motion. In the COM frame, using the notation of Refs. [10, 11] it is,

$$\acute{I}^{(2)}_{s,i} = \frac{2M\nu(s_A - s_B)}{\phi_0^{\text{SF}}(3+2\omega_0)}\left(-\frac{G_{AB}M}{r^2}n^i_{AB}\right), \tag{3.11}$$

where s_A and s_B are the sensitivity of the two bodies.

As the nonlocal contribution starts at 3PN order, the ordinary Lagrangian is

$$L^{\text{tail}}_{\text{ord}} = L^{\text{tail}} + \sum_{J=A,B} m_J\left(-a^i_J - \sum_{J\neq K}\frac{G_{AB}\,m_K}{r^2}n^i_{JK}\right)\xi_{J,i}, \tag{3.12}$$

where $L^{\text{tail}}_{\text{ord}}$ is given by the same expression as Eq. (3.10) but with second-time derivatives of the dipole moment replaced by its on-shell value given in Eq. (3.11), and the nonlocal coordinate shift $\xi_{J,j}$,

$$\xi_{J,j} = \frac{1}{m_J}\frac{2G^2M}{3c^6}(3+2\omega_0)\left[-\frac{m_J(1-2s_J)}{\phi_0^{\text{SF}}(3+2\omega_0)}\right]\delta^i_j \text{Pf}_{2r/c}\int_{-\infty}^{\infty}\frac{d\tau}{|\tau|}\acute{I}^{(2)}_{s,i}(t+\tau). \tag{3.13}$$

The ordinary Hamiltonian is then derived using the ordinary Legendre transformation, $H_{\text{ord}} = \sum_A p_A v_A - L_{\text{ord}}$ which reads $H_{\text{ord}} = H^{\text{loc}}_{\text{ord}} + H^{\text{tail}}_{\text{ord}}$, where the local contribution $H^{\text{loc}}_{\text{ord}}$ is derived in Chap. 2 (see, Appendix 2.9) and the tail contribution is

$$H^{\text{tail}}_{\text{ord}} = -\frac{2G^2M}{3c^6}(3+2\omega_0)\,\text{Pf}_{2r/c}\int_{-\infty}^{\infty}\frac{d\tau}{|\tau|}\acute{I}^{(2)}_{s,i}(t)\acute{I}^{(2)}_{s,i}(t+\tau). \tag{3.14}$$

The tail part of the Hamiltonian is thus just minus the tail part of the Lagrangian (3.10) at 3PN order in ST theory.

As shown in Refs. [18, 20] for the non-order-reduced, nonlocal framework the *Noetherian* conserved energy E_{cons} is not given by the Hamiltonian but is $E_{\text{cons}} = H^{\text{tail}}_{\text{ord}} + \delta H$. This additional term δH consists of purely a constant term (DC type) and a time oscillating term with zero average value (AC type) and is the same as given in Eq. (4.10) of Ref. [11].

3.3.2 Order-Reduced Ordinary Hamiltonian

The second method to derive the conserved energy for the tail part is to work in the order-reduced, local framework as shown in Refs. [17, 18] for GR.

The tail part of the Hamiltonian in ST theory is,

$$H^{\text{tail}} = -\frac{2G^2M}{3c^6}(3+2\omega_0)\left[\text{Pf}_{2r/c}\int_{-\infty}^{\infty}\frac{d\tau}{|\tau|}I^{(2)}_{s,i}(t)I^{(2)}_{s,i}(t+\tau) - 2\ln\left(\frac{\hat{r}}{a}\right)I^{(2)}_{s,i}(t)^2\right]. \tag{3.15}$$

As mentioned in Ref. [20], in the action-angle form there should be an additional term (second term in Eq. (3.15)) which is local to account for the dependence of the Hadamard Partie finie function on the radial separation r at time t i.e., $\hat{r} = a(1 - e\cos(u))$ in action-angle variables, where a is the semi-major axis, e is the eccentricity and u is the "eccentric anomaly".

The basic methodology we use to order-reduce the nonlocal dynamics of the above form is based on Refs. [17, 18] for GR, and consists of four main steps: (i) Re-express the Hamiltonian in terms of action-angle variables, (ii)"order-reduce" the nonlocal dependence on action-angle variables, (iii) expand it in powers of eccentricity, and (iv) finally, eliminate the periodic terms in the order-reduced Hamiltonian by a canonical transformation. All of these steps lead to the order-reduced ordinary local Hamiltonian for the tail part in terms of action-angle variables.

Let us consider the expression of the nonlocal-in-time piece of Eq. (3.15), i.e.

$$\mathscr{K}(t,\tau) = I^{(2)}_{s,i}(t)I^{(2)}_{s,i}(t+\tau). \tag{3.16}$$

This form of the nonlocal piece can be order-reduced by expressing the phase variables at the shifted-time $t+\tau$ in terms of the phase-space variables at time t. Therefore, we require an explicit way of solving the zeroth-order equations of motion so that they can be used to compute the time-shifted variables in terms of local variables at time t. As the zeroth order equations are Newtonian equations, it will be convenient to use the action-angle form of the Newtonian equations of motion,

$$\begin{aligned} \frac{\partial l}{\partial \hat{t}} &= \frac{\partial H_0}{\partial \mathscr{L}} = \frac{1}{\mathscr{L}^3} = \Omega(\mathscr{L}), & \frac{\partial \mathscr{L}}{\partial \hat{t}} &= \frac{\partial H_0}{\partial l} = 0, \\ \frac{\partial \mathscr{G}}{\partial \hat{t}} &= \frac{\partial H_0}{\partial g} = 0, & \frac{\partial g}{\partial \hat{t}} &= \frac{\partial H_0}{\partial \mathscr{G}} = 0, \end{aligned} \tag{3.17}$$

where $\hat{t} = t/(G_{AB}M)$ is the dimensionless time variable, $(\mathscr{L}, l, \mathscr{G}, g)$ are the action-angle variables. The zeroth-order (Newtonian) Hamiltonian in action-angle variable is $H_0 = -1/(2\mathscr{L}^2)$.

Here, the variable $\mathscr{L}$ is conjugate to the "mean anomaly" l and $\mathscr{G}$ is conjugate to the argument of periastron g. In terms of the Keplerian variables, these are

$$\mathscr{L} = \sqrt{a}, \quad \mathscr{G} = \sqrt{a(1-e^2)}. \tag{3.18}$$

From Eq. (3.17), the variables $\mathscr{L}, \mathscr{G}$ and g are independent of time, and l varies linearly with time, hence it will be sufficient to use

$$l(t+\tau) = l(t) + \Omega\,\hat{\tau}, \tag{3.19}$$

where $\hat{\tau} = \tau/(G_{AB}M)$. The order-reduced nonlocal-in-time expression of Eq. (3.16) becomes

$$\mathcal{K}(t,\tau) = \left(\frac{1}{G_{AB}M}\right)^4 \mathcal{K}(\hat{t},\hat{\tau}) = \left(\frac{\Omega}{G_{AB}M}\right)^4 \frac{d^2}{dl^2} I_{s,i}(l) \frac{d^2}{dl^2} I_{s,i}(l+\Omega\hat{\tau}). \tag{3.20}$$

Using the Fourier decomposition of the dipole moment given in Eq. (3.53), we find the structure of the nonlocal-in-time expression $\mathcal{K}(t,\hat{\tau})$ and hence the Hamiltonian. As shown in [17] for GR, all the periodically varying terms can be eliminated by a suitable canonical transformation. Hence, the order-reduced Hamiltonian can be further simplified by replacing H^{tail} with its l-averaged value

$$\bar{H}_{\text{tail}} = \int_0^{2\pi} dl H^{\text{tail}}. \tag{3.21}$$

Using the result

$$\text{Pf}_T \int_0^{\infty} \frac{dv}{v} \cos(\omega v) = -(\gamma_{\text{E}} + \ln(\omega T)) \qquad \forall\, (\omega > 0), \tag{3.22}$$

where γ_{E} is Euler's constant, and inserting the expression of r from Eq. (3.49), the Hamiltonian Eq. (3.21) reads

$$\bar{H}_{\text{tail}} = \frac{8G^2M}{3}\left(\frac{\Omega}{G_{AB}M}\right)^4 (3+2w_0) \sum_{p=1}^{\infty} p^4 |I_{s,i}(p)|^2 \ln\left(\text{e}^{\gamma_{\text{E}}} \frac{2p\,a\,\Omega}{c}\right). \tag{3.23}$$

Now, inserting the Fourier-Bessel expansion of the scalar dipole moment[1] from Eqs. (3.55)–(3.56) (see Appendix 3.6 for the derivation) in Eq. (3.23), the real two-body nonlocal-in-time Hamiltonian in the order-reduced, the local framework is (using notations of Chap. 2)

$$\hat{\bar{H}}_{\text{tail}} \equiv \frac{\bar{H}_{\text{tail}}}{\mu} = \frac{2\nu}{3a^4}\left(2\delta_+ + \frac{\bar{\gamma}_{AB}(\bar{\gamma}_{AB}+2)}{2}\right) \sum_{p=1}^{\infty} \frac{p^2}{e^2} \left\{4e^2 J_{p-1}^2(pe) + (8-4e^2) J_p^2(pe) - 8e J_{p-1}(pe) J_p(pe)\right\} \left[\gamma_{\text{E}} + \ln\left(\frac{2p\,a^{-1/2}}{c}\right)\right]. \tag{3.24}$$

Expanding the result in powers of the eccentricity, the Hamiltonian as an expansion in eccentricity up to the order of e^4 reads

[1] It should be noted that expressing the scalar dipole moments as Fourier-Bessel expansion does not impact the regime of validity of PN results. Moreover, the Fourier-Bessel expansion considered in this chapter is valid for bound orbits only and the expansion for unbound orbits is given in Chap. 4.

$$\hat{\bar{H}}_{\text{tail}} = \frac{2\nu}{3a^4}\left(2\delta_+ + \frac{\bar{\gamma}_{AB}(\bar{\gamma}_{AB}+2)}{2}\right)\Big\{2\ln(2) - \ln(a) + 2\gamma_{\text{E}} + e^2\left(14\ln(2) + 6\gamma_{\text{E}} - 3\ln(a)\right)$$
$$+ e^4\left(\frac{45}{4}\gamma_{\text{E}} - \frac{3}{4}\ln(2) + \frac{729}{32}\ln(3) - \frac{45}{8}\ln(a)\right) + \mathcal{O}(e^6)\Big\}. \tag{3.25}$$

3.4 Scalar Tensor Corrections to Effective-One-Body Hamiltonian at 3PN:Tail

In this section, we will derive the complete tail corrections to the EOB metric potentials (A, B, Q_e) for ST theories at 3PN order.

Similar to the decomposition of the complete 3PN coefficient δa_4^{ST} in Eq. (2.45) of Chap. 2, we decompose the complete 3PN ST coefficients δb_3^{ST} and δq_3^{ST} as

$$\delta b_3^{ST} = \delta b_{3,\text{loc}}^{\text{ST}} + \delta b_{3,\text{nonloc}}^{\text{ST}} + \delta b_{3,\text{tidal}}^{\text{ST}}, \tag{3.26}$$
$$\delta q_3^{ST} = \delta q_{3,\text{loc}}^{\text{ST}} + \delta q_{3,\text{nonloc}}^{\text{ST}} + \delta q_{3,\text{tidal}}^{\text{ST}}, \tag{3.27}$$

where the local contributions $(\delta b_{3,\text{loc}}^{\text{ST}}, \delta q_{3,\text{loc}}^{\text{ST}})$ are derived in Chap. 2 (see Eqs. (2.37) and (2.38)), $(\delta b_{3,\text{nonloc}}^{\text{ST}}, \delta q_{3,\text{nonloc}}^{\text{ST}})$ are the nonlocal contributions, and $(\delta b_{3,\text{tidal}}^{\text{ST}}, \delta q_{3,\text{tidal}}^{\text{ST}})$ are the tidal contributions. The nonlocal contributions can be further decomposed similarly to Eq. (2.45) of Chap. 2 as

$$\delta a_{4,\text{nonloc}}^{\text{ST}} = \delta a_{4,\text{nonloc},0}^{\text{ST}} + \delta a_{4,\text{nonloc},\log}^{\text{ST}}\ln(\hat{r}), \tag{3.28}$$
$$\delta b_{3,\text{nonloc}}^{\text{ST}} = \delta b_{3,\text{nonloc},0}^{\text{ST}} + \delta b_{3,\text{nonloc},\log}^{\text{ST}}\ln(\hat{r}), \tag{3.29}$$
$$\delta q_{3,\text{nonloc}}^{\text{ST}} = \delta q_{3,\text{nonloc},0}^{\text{ST}} + \delta q_{3,\text{nonloc},\log}^{\text{ST}}\ln(\hat{r}). \tag{3.30}$$

Inserting the split of the EOB functions (A, B, q_3) using Eqs. (3.26)–(3.27) and (2.45) of Chap. 2 into the effective Hamiltonian of Eq. (3.2) and expanding the right-hand side into a Taylor series in $1/c^2$, we obtain

$$\hat{H}_{\text{eff}} = \hat{H}_{\text{eff}}^{\text{loc}} + \hat{H}_{\text{eff}}^{\text{nonloc}}, \tag{3.31}$$

where $\hat{H}_{\text{eff}}^{\text{loc}}$ is computed only from the local contributions $(\delta a_{4,\text{loc}}^{\text{ST}}, \delta b_{3,\text{loc}}^{\text{ST}}, \delta q_{3,\text{loc}}^{\text{ST}})$ and $\hat{H}_{\text{eff}}^{\text{nonloc}}$ is the nonlocal contribution of the Hamiltonian computed from $(\delta a_{4,\text{nonloc}}^{\text{ST}}, \delta b_{3,\text{nonloc}}^{\text{ST}}, \delta q_{3,\text{nonloc}}^{\text{ST}})$. The nonlocal contribution $\hat{H}_{\text{eff}}^{\text{nonloc}}$ reads

$$\hat{H}_{\text{eff}}^{\text{nonloc}} = \frac{1}{2}\left(\delta a_{4,\text{nonloc}}^{\text{ST}}\frac{1}{\hat{r}^4} - \delta b_{3,\text{nonloc}}^{\text{ST}}\frac{\hat{p}_r^2}{\hat{r}^3} + \delta q_{3,\text{nonloc}}^{\text{ST}}\frac{\hat{p}_r^4}{\hat{r}^2}\right). \tag{3.32}$$

To map the real two-body dynamics to EOB, we express the nonlocal effective Hamiltonian, $\hat{H}_{\text{eff}}^{\text{nonloc}}$, in action-angle variables $\mathcal{L}, l, \mathcal{G}$, and g (hence the Keplerian variables a and e) and compute its l-averaged value,

$$\hat{\bar{H}}_{\rm eff}^{\rm nonloc} = \frac{1}{2\pi}\int_0^{2\pi} dl\, \hat{H}_{\rm eff}^{\rm nonloc}. \tag{3.33}$$

The explicit expression of $\hat{H}_{\rm eff}^{\rm nonloc}$ depends on the l-averaged monomials involving powers of $1/\hat{r}$ and $\hat{p}_r$ (and also $\ln(\hat{r})$ from Eqs. (3.28), (3.29), and (3.30)). These computations can be performed by expanding Eq. (3.32) in terms of the eccentricity up to e^5 using the Newtonian equations of motion in the action-angle form recalled in Sect. 3.3.2. The l-averaged value we obtain is

$$\begin{aligned}
\hat{\bar{H}}_{\rm eff}^{\rm nonloc} &= \frac{1}{2a^4}\Big\{\delta a_{4,\rm nonloc,0}^{\rm ST} + \delta a_{4,\rm nonloc,log}^{\rm ST}\ln(a)\\
&+ \left(3\delta a_{4,\rm nonloc,0}^{\rm ST} - \frac{7}{4}\delta a_{4,\rm nonloc,log}^{\rm ST} - \frac{1}{2}\delta b_{3,\rm nonloc,0}^{\rm ST} + 3\delta a_{4,\rm nonloc,log}^{\rm ST}\ln(a) - \frac{1}{2}\delta b_{3,\rm nonloc,log}^{\rm ST}\ln(a)\right)e^2\\
&+ \left(\frac{45}{8}\left[\delta a_{4,\rm nonloc,0}^{\rm ST} + \delta a_{4,\rm nonloc,log}^{\rm ST}\ln(a)\right] - \frac{5}{4}\left[\delta b_{3,\rm nonloc,0}^{\rm ST} + \delta b_{3,\rm nonloc,log}^{\rm ST}\ln(a)\right]\right.\\
&\left.+\frac{3}{8}\left[\delta q_{3,\rm nonloc,0}^{\rm ST} + \delta q_{3,\rm nonloc,log}^{\rm ST}\ln(a)\right] - \frac{171}{32}\delta a_{4,\rm nonloc,log}^{\rm ST} + \frac{9}{16}\delta b_{3,\rm nonloc,log}^{\rm ST}\right)e^4 + \mathcal{O}(e^6)\Big\}.
\end{aligned} \tag{3.34}$$

The final step is then to map the real two-body dynamics to the EOB metric by the *nontrivial* map

$$\hat{H}_{\rm real} = \frac{H_{\rm real}}{\mu} = \frac{1}{\nu}\sqrt{1 + 2\nu(\hat{H}_{\rm eff} - 1)}, \tag{3.35}$$

between the EOB Hamiltonian ($\hat{H}_{\rm eff}$) and the real two-body Hamiltonian ($\hat{H}_{\rm real}$). The quadratic map relating the two Hamiltonians has been proved at *all* PN orders in GR and ST within the Post-Minkowskian scheme in Ref. [22]. However, it can be seen that only for the nonlocal contributions at 3PN order, the map relating the two nonlocal Hamiltonians is

$$\hat{\bar{H}}_{\rm eff}^{\rm nonloc} = \hat{\bar{H}}_{\rm real,nonloc}^{\rm II}, \tag{3.36}$$

where $\hat{\bar{H}}_{\rm real,nonloc}^{\rm II} = \hat{\bar{H}}_{\rm tail}$. The *unique* nonlocal ST contributions at 3PN from this matching are

$$\delta a_{4,\rm nonloc,0}^{\rm ST} = \frac{4}{3}\nu\left[2\delta_+ + \frac{\bar{\gamma}_{AB}(\bar{\gamma}_{AB}+2)}{2}\right](2\ln 2 + 2\gamma_{\rm E}), \tag{3.37}$$

$$\delta a_{4,\rm nonloc,log}^{\rm ST} = -\frac{4}{3}\nu\left[2\delta_+ + \frac{\bar{\gamma}_{AB}(\bar{\gamma}_{AB}+2)}{2}\right], \tag{3.38}$$

$$\delta b_{3,\rm nonloc,0}^{\rm ST} = \frac{4}{3}\nu\left[2\delta_+ + \frac{\bar{\gamma}_{AB}(\bar{\gamma}_{AB}+2)}{2}\right]\left(\frac{21}{2} - 16\ln 2\right), \tag{3.39}$$

$$\delta b_{3,\rm nonloc,log}^{\rm ST} = 0, \tag{3.40}$$

$$\delta q^{\rm ST}_{3,\rm nonloc,0} = \frac{4}{3}\nu\left[2\delta_+ + \frac{\bar{\gamma}_{AB}(\bar{\gamma}_{AB}+2)}{2}\right]\left(-\frac{31}{4} - \frac{256}{3}\ln 2 + \frac{243}{4}\ln 3\right), \tag{3.41}$$

$$\delta q^{\rm ST}_{3,\rm nonloc,log} = 0. \tag{3.42}$$

The ST tensor correction $\delta a^{\rm ST}_{4,\rm nonloc}$ for the circular orbit case, Eqs. (3.37)–(3.38), matches with the results obtained in Chap. 2 (see, Eqs. (2.47)–(2.48)) using results of Ref. [11] except with a negative sign in Eq. (3.38). The negative sign is due to the difference in the definition of $\delta a^{\rm ST}_{4,\rm nonloc}$ in Eq. (3.28) used in this Chapter compared to the Eq. (2.45) of Chap. 2.

3.5 Conclusions

In Chap. 2, building upon the results of [11] for massless scalar-tensor theory, we have determined the EOB coefficients at 3PN order though restricting ourselves to the local-in-time part of the dynamics and nonlocal-in-time and tail contributions only for the circular case. In the present chapter, we have derived the complete nonlocal-in-time EOB coefficients starting from the nonlocal-in-time Lagrangian of Ref. [11]. First, we have derived the two-body *conserved* ordinary Hamiltonian (dependent only on positions and momenta) for a nonlocal-in-time part by two methods: (i) non-order-reduced nonlocal Hamiltonian using nonlocal shift (see, Refs. [19, 20] for GR), and (ii) order-reduction of nonlocal dynamics to the local ordinary action-angle Hamiltonian [17]. We have then expressed the effective Hamiltonian in action-angle variables to recast the order-reduced ordinary action-angle Hamiltonian in terms of an equivalent, 3PN-accurate, nonlocal part of the EOB potentials (A, B, Q_e), see Eqs. (3.37)–(3.42).

By combining the results of Chap. 2 and the present Chapter, we could transcribe the two-body Hamiltonian into equivalent 3PN-accurate EOB potentials (A, B, Q_e) for both local-in-time and nonlocal-in-time parts of the dynamics. During the preparation of Ref. [1], we became aware of an independent effort computing similar contributions [23]. We have cross-checked our results with results presented in Ref. [23], specifically the complete (including both local and non-local) correction in A, B, and Q_e EOB potentials. Although computed following different steps, we found that the two results are consistent.

3.6 Fourier Coefficients of the Dipole Moment in Scalar-Tensor Theory

In this Appendix, we will determine the explicit expressions of the *Newtonian* dipole moment in ST theory using the known Fourier decomposition of the Keplerian motion (see Refs. [24, 25] for GR).

The dipole moment, $I_{s,i}(t)$, in the COM frame is

$$I_{s,i}(t) = \frac{2M\nu(s_A - s_B)}{\phi_0^{\mathrm{SF}}(3 + 2w_0)} \, \mathrm{x}_i, \tag{3.43}$$

where $\mathrm{x}_i = (Z_A - Z_B)_i$ is the relative separation vector and $Z_{A,B}$ indicate the positions of the two bodies.

Since the motion is planar, we can choose the coordinate system (x, y, z) such that the motion takes place in the xy-plane. Using polar coordinates $(\hat{r}, \phi_a)$ we obtain

$$x = \hat{r}\ \cos(\phi_a), \quad y = \hat{r}\ \sin(\phi_a), \tag{3.44}$$

where (x, y) are the coordinates of the dimensionless relative separation, $\hat{r} = x_A - x_B$ with $x_J = \mathrm{x}_J/(G_{AB}M)$ denoting the position of two bodies.

As mentioned in Refs. [17, 24, 25] for GR, for leading order contributions it is convenient to use the Delaunay (action-angle) form of the Newtonian equations of motion. In terms of the action-angle variables $(\mathscr{L}, l, \mathscr{G}, g)$, the Cartesian coordinates (x, y) are given by (Here, following the notations of [26])

$$x = x_0 \cos(g) - y_0 \sin(g), \tag{3.45}$$
$$y = x_0 \cos(g) + y_0 \sin(g), \tag{3.46}$$
$$x_0 = \hat{r} \cos(f) = a(\cos(u) - e), \tag{3.47}$$
$$y_0 = \hat{r} \sin(f) = a\sqrt{1 - e^2} \sin(u), \tag{3.48}$$
$$\hat{r} = a(1 - e\cos(u)), \tag{3.49}$$

where a is the semi-major axis, e is the eccentricity, f is the "true anamoly" and the "eccenteric anamoly" u in terms of Bessels functions is given by

$$u = l + \sum_{n=1}^{\infty} \frac{2}{n} \mathrm{J}_n(ne) \sin(nl). \tag{3.50}$$

The Bessel-Fourier expansion of $\cos(u)$ and $\sin(u)$, which directly enters x_0, y_0 are:

$$\cos(u) = -\frac{e}{2} + \sum_{n=1}^{\infty} \frac{1}{n} \left[\mathrm{J}_{n-1}(ne) - \mathrm{J}_{n+1}(ne)\right] \cos(nl), \tag{3.51}$$

$$\sin(u) = \sum_{n=1}^{\infty} \frac{1}{n} \left[\mathrm{J}_{n-1}(ne) + \mathrm{J}_{n+1}(ne)\right] \sin(nl). \tag{3.52}$$

From Eqs. (3.45)–(3.50), the dipole moment $I_{s,i}$ is a periodic function of l (and hence time) at the Newtonian order. Thus it can be decomposed into the Fourier series

$$I_{s,i}(l) = \sum_{p=-\infty}^{\infty} I_{i,s}(p)\, e^{ipl}, \tag{3.53}$$

with

$$I_{s,i}(p) = \frac{1}{2\pi}\int_0^{2\pi} dl\; I_{s,i}\, e^{-ipl}. \tag{3.54}$$

The Fourier coefficients of the scalar dipole moment at Newtonian order are derived using Eq. (3.54) in terms of combinations of Bessel Functions.

Inserting the expression of Cartesian coordinates in terms of action-angle variables using Eqs. (3.45)–(3.52), we find the Fourier-Bessel coefficients of the scalar dipole moment are,

$$\begin{aligned} I_{s,x}(p) = \; & G_{AB}M\left[\frac{2M\nu(s_A - s_B)}{\phi_0^{\mathrm{SF}}(3+2w_0)}\frac{a}{2p}\left\{\left[J_{p-1}(pe) - J_{p+1}(pe)\right]\cos(g)\right.\right. \\ & \left.\left. + i\sqrt{1-e^2}\left[J_{p-1}(pe) + J_{p+1}(pe)\right]\sin(g)\right\}\right], \end{aligned} \tag{3.55}$$

$$\begin{aligned} I_{s,y}(p) = \; & G_{AB}M\left[\frac{2M\nu(s_A - s_B)}{\phi_0^{\mathrm{SF}}(3+2w_0)}\frac{a}{2p}\left\{\left[J_{p-1}(pe) - J_{p+1}(pe)\right]\sin(g)\right.\right. \\ & \left.\left. + i\sqrt{1-e^2}\left[J_{p-1}(pe) + J_{p+1}(pe)\right]\cos(g)\right\}\right]. \end{aligned} \tag{3.56}$$

References

1. Jain T (2023) Nonlocal-in-time effective one body Hamiltonian in scalar-tensor gravity at third post-Newtonian order. Phys Rev D 107:084018. arXiv:2301.01070. https://doi.org/10.1103/PhysRevD.107.084018 [gr-qc]
2. Damour T, Esposito-Farese G (1992) Tensor multiscalar theories of gravitation. Class Quant Grav 9:2093–2176. https://doi.org/10.1088/0264-9381/9/9/015
3. Damour T, Esposito-Farese G (1993) Nonperturbative strong field effects in tensor - scalar theories of gravitation. Phys Rev Lett 70:2220–2223. https://doi.org/10.1103/PhysRevLett.70.2220
4. Damour T, Esposito-Farese G (1996) Testing gravity to second postNewtonian order: a field theory approach. Phys Rev D 53:5541–5578. arXiv:gr-qc/9506063. https://doi.org/10.1103/PhysRevD.53.5541
5. Freire PCC, Wex N, Esposito-Farèse G, Verbiest JP, Bailes M, Jacoby BA, Kramer M, Stairs IH, Antoniadis J, Janssen GH (2012) The relativistic pulsar-white dwarf binary PSR J1738+0333 II. The most stringent test of scalar-tensor gravity. Mon Not Roy Astron Soc 423:3328. arXiv:1205.1450. https://doi.org/10.1111/j.1365-2966.2012.21253.x [astro-ph.GA]
6. Khalil M, Mendes RFP, Ortiz N, Steinhoff J (2022) Effective-action model for dynamical scalarization beyond the adiabatic approximation. arXiv:2206.13233 [gr-qc]
7. Gautam T, Freire PCC, Batrakov A, Kramer M, Miao CC, Parent E, Zhu WW (2022) Relativistic effects in a mildly recycled pulsar binary: PSR J1952+2630. arXiv:2210.03464 [astro-ph.HE]

8. Palenzuela C, Barausse E, Ponce M, Lehner L (2014) Dynamical scalarization of neutron stars in scalar-tensor gravity theories. Phys Rev D 89:044024. arXiv:1310.4481. https://doi.org/10.1103/PhysRevD.89.044024 [gr-qc]
9. Sennett N, Buonanno A (2016) Modeling dynamical scalarization with a resummed post-Newtonian expansion. Phys Rev D 93:124004. arXiv:1603.03300. https://doi.org/10.1103/PhysRevD.93.124004 [gr-qc]
10. Bernard L (2018) Dynamics of compact binary systems in scalar-tensor theories: equations of motion to the third post-Newtonian order. Phys Rev D 98:044004. arXiv:1802.10201. https://doi.org/10.1103/PhysRevD.98.044004 [gr-qc]
11. Bernard L (2019) Dynamics of compact binary systems in scalar-tensor theories: II. Center-of-mass and conserved quantities to 3PN order. Phys Rev D 99:044047. arXiv:1812.04169. https://doi.org/10.1103/PhysRevD.99.044047 [gr-qc]
12. Buonanno A, Damour T (1999) Effective one-body approach to general relativistic two-body dynamics. Phys Rev D59:084006. arXiv:gr-qc/9811091. https://doi.org/10.1103/PhysRevD.59.084006
13. Buonanno A, Damour T (2000) Transition from inspiral to plunge in binary black hole coalescences. Phys Rev D62:064015. arXiv:gr-qc/0001013. https://doi.org/10.1103/PhysRevD.62.064015
14. Damour T, Jaranowski P, Schaefer G (2000) On the determination of the last stable orbit for circular general relativistic binaries at the third postNewtonian approximation. Phys Rev D62:084011. arXiv:gr-qc/0005034. https://doi.org/10.1103/PhysRevD.62.084011 [gr-qc]
15. Damour T, Jaranowski P, Schäfer G (2008) Effective one body approach to the dynamics of two spinning black holes with next-to-leading order spin-orbit coupling. Phys Rev D78:024009. arXiv:0803.0915. https://doi.org/10.1103/PhysRevD.78.024009 [gr-qc]
16. Damour T, Jaranowski P, Schäfer G (2014) Nonlocal-in-time action for the fourth post-Newtonian conservative dynamics of two-body systems. Phys Rev D 89:064058. arXiv:1401.4548. https://doi.org/10.1103/PhysRevD.89.064058
17. Damour T, Jaranowski P, Schäfer G (2015) Fourth post-Newtonian effective one-body dynamics. Phys Rev D 91:084024. arXiv:1502.07245. https://doi.org/10.1103/PhysRevD.91.084024 [gr-qc]
18. Damour T, Jaranowski P, Schäfer G (2016) Conservative dynamics of two-body systems at the fourth post-Newtonian approximation of general relativity. Phys Rev D93:084014. arXiv:1601.01283. https://doi.org/10.1103/PhysRevD.93.084014 [gr-qc]
19. Bernard L, Blanchet L, BohÃ A, Faye G, Marsat S (2016) Fokker action of nonspinning compact binaries at the fourth post-Newtonian approximation. Phys Rev D93:084037. arXiv:1512.02876. https://doi.org/10.1103/PhysRevD.93.084037 [gr-qc]
20. Bernard L, Blanchet L, Bohé A, Faye G, Marsat S (2017) Energy and periastron advance of compact binaries on circular orbits at the fourth post-Newtonian order. Phys Rev D95:044026. arXiv:1610.07934. https://doi.org/10.1103/PhysRevD.95.044026 [gr-qc]
21. Damour T, Jaranowski P, Schaefer G (2000) Dynamical invariants for general relativistic two-body systems at the third postNewtonian approximation. Phys Rev D 62:044024. arXiv:gr-qc/9912092. https://doi.org/10.1103/PhysRevD.62.044024
22. Damour T (2016) Gravitational scattering, post-Minkowskian approximation and effective one-body theory. Phys Rev D94:104015. arXiv:1609.00354. https://doi.org/10.1103/PhysRevD.94.104015 [gr-qc]
23. Julié F-L, Baibhav V, Berti E, Buonanno A (2023) Third post-Newtonian effective-one-body Hamiltonian in scalar-tensor and Einstein-scalar-Gauss-Bonnet gravity. Phys Rev D 107:104044. arXiv:2212.13802. https://doi.org/10.1103/PhysRevD.107.104044 [gr-qc]
24. Peters PC, Mathews J (1963) Gravitational radiation from point masses in a Keplerian orbit. Phys Rev 131:435–439. https://doi.org/10.1103/PhysRev.131.435
25. Arun KG, Blanchet L, Iyer BR, Qusailah MSS (2008) Tail effects in the 3PN gravitational wave energy flux of compact binaries in quasi-elliptical orbits. Phys Rev D 77:064034. arXiv:0711.0250. https://doi.org/10.1103/PhysRevD.77.064034 [gr-qc]
26. Brouwer D, Clemence GM (1961) Methods of celestial mechanics. Academic Press

Chapter 4
Gravitational Scattering up to Third Post-Newtonian Order for Conservative Dynamics in Scalar-Tensor Theories

We compute the scattering angle χ for hyperbolic-like encounters in massless Scalar-Tensor (ST) theories up to 3PN order for the conservative part of the dynamics. To calculate the gauge-invariant scattering angle as a function of energy and orbital angular momentum, we use the approach of Effective-One-Body (EOB) Hamiltonian. We compute the nonlocal-in-time contribution to the scattering angle by using the strategy of order-reduction of nonlocal dynamics introduced for small-eccentricity orbits.

With the future improvements in the sensitivity of the ground-based detectors and with the possibility of next generation detectors, the detector frequency band will observe more cycles during the inspiral phase which, as mentioned previously, is studied using analytical models such as the PN approximation. Therefore, to exploit the physics from future observations it is important to improve the analytical information at higher PN orders. For this purpose, there is a considerable interest in using the ideas from particle physics to compute the scattering angle for classical two-body problem (see, Ref. [1] for early ideas to map scattering information to classical physics). This is because the effective field theory (EFT) approach introduced in Ref. [2] to compute the scattering angle reduces the classical two-body problem into the computation of Feynman diagrams which provides a more powerful tool to extend the dynamics to higher PN orders. Moreover, the EFT approach has been extended to post-Minkowskian (PM) framework [3] in which the scattering angle is computed perturbatively in powers of the gravitational constant G to all orders in the velocity. This implies that each PM order encapsulates the information contained at all the PN orders.

Now, at first glance, it may seem that the scattering angle computed for the unbound orbits (hyperbolic encounters) is unrelated to the bound orbit binary systems observed using GWs. However, a remarkable connection has been developed to relate the scattering angle to the observable dynamical invariants like the orbital frequency of the bound orbit binary system in the form of a 'Boundary-to-Bound' dictionary [4]. Furthermore, the connection between the scattering information and the bound orbit binary dynamics was also emphasised in Ref. [5] in the context of the EOB approach.

T. Jain, *Surfing the Gravitational Waves*, Springer Theses,
https://doi.org/10.1007/978-3-031-96180-9_4

In this work, we use the EOB approach to compute the scattering angle for hyperbolic-like encounters for ST theories. In future, the results computed here can act as a computational check for the scattering angle computed using the EFT machinery.

Additionally, it has been pointed out in Ref. [6] that the hyperbolic encounters for which the primary sources are dynamical capture scenarios [7] in globular clusters, maybe detected by the current ground-based GW detectors. Therefore, analytically studying the scattering angle for the classical two-body problem for hyperbolic encounters is important in modified theories of gravity. Since, in this thesis, we focus on ST theories, we compute the scattering angle using the EOB approach in ST theories.

This is based on the work of Ref. [8].

4.1 Introduction

The EOB description of the unbound, scattering states of binary systems was introduced in Ref. [5]. Recently, the approach has been used to compute the scattering angle within the PN approximation in GR [9]. The main aim of this work is to compute the scattering angle in ST theories up to 3PN order using the ST EOB Hamiltonian derived in Chaps. 2 and 3.

The chapter is organised as follows. In Sect. 4.2, we briefly summarise the ST theory parameters used and the structure of the EOB metric potentials in ST theories up to 3PN order. Then, in Sect. 4.3 we derive the scattering angle for the local part of the dynamics in ST theories at the 3PN order and derive the scattering angle for the nonlocal part of the dynamics at 3PN order using the order-reduction approach in Sect. 4.4. Finally, in Sect. 4.5 we sum the local and nonlocal contribution at 3PN in large angular momentum expansion. From now on, we use geometric units with $c = G = 1$ unless specified explicitly.

4.2 Scalar-Tensor Parameters and Effective-One-Body Formalism

As in the previous chapters, here also, we consider massless ST theories with single scalar field described by the minimal coupling of the scalar field to the metric in the Einstein Frame, and its action reads

$$S = \frac{c^4}{16\pi G}\int d^4x\sqrt{-g}(R - 2g^{\mu\nu}\partial_\mu\varphi\partial_\nu\varphi) + S_m[\Psi, \mathscr{A}(\varphi)^2 g_{\mu\nu}], \tag{4.1}$$

where $g_{\mu\nu}$ is the Einstein metric, R is the Ricci scalar, φ is the scalar field, Ψ collectively denotes the matter fields, $g \equiv \det(g_{\mu\nu})$ and G is the Newton's gravitational constant.

Here also, the dimensionless body-dependent parameters that encompass the scalar field effect in ST theories are expressed in DEF conventions following Refs. [10, 11], i.e. the Einstein frame. The conversion of the Jordan Frame parameters to DEF conventions is given in Table 2.1 of Chap. 2.

As mentioned in Chap. 2, the generalisation of the EOB formalism to 3PN order introduces non-geodesic contributions (Q_e) to the effective dynamics (see, Ref. [12]). The structure of the non-geodesic term Q_e at 3PN is given in Eq. (2.26). Similar to Chap. 2, here also, we use the gauge freedom introduced in Ref. [12], known as DJS gauge, to set $q_1 = q_2 = 0$ so that Q_e only depends on the radial momentum.

The relation between the real and EOB Hamiltonian used in Chaps. 2 and 3 for the bound orbits (elliptic-like),

$$\hat{H}_{\text{real}}(\mathbf{r}, \mathbf{p}) \equiv \frac{H_{\text{real}}}{\mu} = \frac{1}{\nu}\sqrt{1 + 2\nu\left(\hat{H}_{\text{eff}} - 1\right)}, \tag{4.2}$$

hold for the unbound orbits (hyperbolic-like) considered in this Chapter (see Ref. [5]) also. Here also, M, μ and ν are the total mass, the reduced mass and the symmetric mass ratio, respectively.

The dimensionless variables used in this Chapter are defined as,

$$\hat{p}_r = \frac{p_r}{\mu}; \quad j = \hat{p}_\phi = \frac{p_\phi}{G_{AB} M \mu}; \quad \hat{r} = \frac{r}{G_{AB} M}. \tag{4.3}$$

Hereafter, *hat* will be used to denote the dimensionless variables.

The three EOB potentials (A, B, Q_e) up to 3PN order in DJS gauge choice of Ref. [12] formally read

$$A(\hat{r}) = 1 - \frac{2}{\hat{r}} + \frac{a_2(\nu)}{\hat{r}^2} + \frac{a_3(\nu)}{\hat{r}^3} + \frac{a_4(\nu)}{\hat{r}^4}, \tag{4.4}$$

$$B(\hat{r}) = 1 + \frac{b_1(\nu)}{\hat{r}} + \frac{b_2(\nu)}{\hat{r}^2} + \frac{b_3(\nu)}{\hat{r}^3}, \tag{4.5}$$

$$\hat{Q}_e(\hat{r}) = q_3(\nu)\frac{\hat{p}_r^4}{\hat{r}^2}, \tag{4.6}$$

where the ν-dependent coefficients a_i, b_i, and q_i take into account both GR and ST corrections which are separated as

$$a_i = a_i^{\text{GR}} + a_{i,\text{ST}}, \tag{4.7}$$

$$b_i = b_i^{\text{GR}} + b_{i,\text{ST}}, \tag{4.8}$$

$$q_3 = q_3^{\text{GR}} + q_{3,\text{ST}}. \tag{4.9}$$

As mentioned previously, for ST theories the nonlocal-in-time and tidal contributions also start at the 3PN-order [13, 14], therefore, the ST corrections to EOB potentials can be further decomposed as Eq. (2.45). However, in this Chapter, for the sake of simplicity of solving tail contributions, we decompose the metric potentials as Ref. [15]

$$a_{4,\mathrm{ST}} = a^{\mathrm{I}}_{4,\mathrm{ST}} + a^{\mathrm{II}}_{4,\mathrm{ST}}, \tag{4.10}$$

$$b_{3,\mathrm{ST}} = b^{\mathrm{I}}_{3,\mathrm{ST}} + b^{\mathrm{II}}_{3,\mathrm{ST}}, \tag{4.11}$$

$$q_{3,\mathrm{ST}} = q^{\mathrm{I}}_{3,\mathrm{ST}} + q^{\mathrm{II}}_{3,\mathrm{ST}}, \tag{4.12}$$

where the superscripts I and II denote the local and nonlocal contributions, respectively. These coefficients are then further decomposed as,

$$a^{\mathrm{I}}_{4,\mathrm{ST}} = a^{\mathrm{loc}}_{4,\mathrm{ST}} + a^{\mathrm{log}}_{4,\mathrm{ST}} \ln(u), \tag{4.13}$$

$$b^{\mathrm{I}}_{3,\mathrm{ST}} = b^{\mathrm{loc}}_{3,\mathrm{ST}} + b^{\mathrm{log}}_{3\mathrm{ST}} \ln(u), \tag{4.14}$$

$$q^{\mathrm{I}}_{3,\mathrm{ST}} = q^{\mathrm{loc}}_{3,\mathrm{ST}} + q^{\mathrm{log}}_{3,\mathrm{ST}} \ln(u), \tag{4.15}$$

where $u = 1/\hat{r}$.

These corrections to the EOB potentials in ST theories up to 3PN order have been derived in Refs. [16–20] based on the real two-body Lagrangian up to 3PN order given in Refs. [13, 14, 21, 22].

4.3 Scalar-Tensor Scattering Angle: Local Contributions

In this section, we derive the contribution to the scattering angle for encounters of two non-spinning bodies for the local part of the conservative dynamics up to 3PN order in ST theories. As the nonlocal-in-time (tail) effects start only at the 3PN level in ST theories, the scattering angle up to 3PN can be separated as a sum of functions,

$$\chi = \chi_{\mathrm{loc}} + \chi_{\mathrm{nonloc}}, \tag{4.16}$$

where χ_{loc} and χ_{nonloc} are, respectively, local and nonlocal contributions to the scattering angle.

The convenient way to compute the scattering angle is to use the Hamilton-Jacobi approach.[1] For ST theories as in GR, the Hamiltonians H_{eff} and H_{real} are invariant

[1] In the Hamiltonian-Jacobi formalism, the solution of the equation of motion proceeds as follows. First, using the Hamiltonian we form the Hamiltonian-Jacobi Equation,

$$\frac{\partial S}{\partial t} + H\left(q; \frac{\partial S}{\partial q}; t\right) = 0, \tag{4.17}$$

and find the complete integral of motion $S(q, \alpha_c, t)$, where α_c is an arbitrary constant. Then by differentiating $S(q, \alpha_c, t)$ with respect to alpha and equating this equation with a constant, we find the solutions of the equations of motion, i.e. the coordinates q. For details refer to Sect. 47 of Ref. [23].

under time-translations and space rotations, and their associated constants of motion are, respectively the energy and the angular momentum of the binary system. The EOB action obtained from separating the EOB coordinates takes the form,

$$S(t_{\text{eff}}, r, \phi; \mathscr{E}_{\text{eff}}, p_\phi) = -\,\mathscr{E}_{\text{eff}} t_{\text{eff}} + p_\phi \phi + \int^r dr\ p_r(r, \mathscr{E}_{\text{eff}}, p_\phi), \tag{4.18}$$

where $\mathscr{E}_{\text{eff}}$ is the energy in the EOB formalism and t_{eff} is the EOB metric coordinate time. Using the Hamiltonian-Jacobi formalism, the solution of the equation of motion for the orbit (ϕ) is obtained from (Sect. 47 of Ref. [23]),

$$\frac{\partial S}{\partial p_\phi} = \phi_0 = \text{const.}, \tag{4.19}$$

and the scattering angle related to the orbital phase is then [9]

$$\frac{1}{2}\left(\chi(\hat{E}, j) + \pi\right) = -\int_0^{u_{(max)}} \frac{1}{u^2}\frac{\partial}{\partial j}\hat{p}_r(\hat{E}, j, u)du, \tag{4.20}$$

where $u = 1/\hat{r}$, $u_{(max)} = 1/r_{min}$ is the distance of the closest approach of the two bodies, j is the dimensionless angular momentum defined in Eq. (4.3), and $\hat{E}$ is the dimensionless energy variable defined as [9]

$$\hat{E} \equiv \frac{1}{2}(\hat{\mathscr{E}}_{\text{eff}}^2 - 1) \equiv \frac{1}{2}p_\infty^2, \tag{4.21}$$

where $\hat{\mathscr{E}}_{\text{eff}} = \mathscr{E}_{\text{eff}}/\mu$ is the dimensionless EOB energy and $p_\infty^2 \equiv 2\hat{E}$ is an energy variable. (The notation p_∞ is introduced instead of the notation v_∞ of Ref. [9] to be consistent with the current literature.)

4.3.1 PN-Expanded χ_{loc} for Scalar-Tensor Theories

Let us now compute the radial momentum $\hat{p}_r$ as a function of $u = 1/\hat{r}$ and orbital angular momentum and energy which would then be used to compute the explicit integral of Eq. (4.20). This is obtained by iteratively solving for $\hat{p}_r^2$ the EOB energy conservation law, $\hat{\mathscr{E}}_{\text{eff}}^2 = \hat{H}_{\text{eff}}^2$ (see, Eq. (3.2)),

$$\hat{\mathscr{E}}_{\text{eff}}^2 = A(u)\left(1 + \frac{\hat{p}_r^2}{B(u)} + j^2u^2 + q_3(\nu)\,\hat{p}_r^4 u^2\right), \tag{4.22}$$

which yields,

$$\hat{p}_r^2(\hat{E}, j, u) = [\hat{p}_r^2]^0 + [\hat{p}_r^2]^1\eta^2 + [\hat{p}_r^2]^2\eta^4 + [\hat{p}_r^2]^3\eta^6 + \mathscr{O}(\eta^8) \tag{4.23}$$

with $\eta \sim 1/c$ as a PN-order marker. The Newtonian order contribution to $\hat{p}_r^2$ is

$$[\hat{p}_r^2]^0 = 2\hat{E} - j^2u^2 + 2u \tag{4.24}$$

The explicit expressions for $\hat{p}_r^2$ and hence, $\hat{p}_r$, up to 3PN order in ST theory are given in the auxiliary file attached with Ref. [8].

This kind of formal PN expansion of $\hat{p}_r$, and, hence, the expansion of the integrand of Eq. (4.20) along with the PN expansion of the function $u_{(\max)}$ generates a sequence of divergent integrals on the range $[0, u_{(\max)}]$. The integral of Eq. (4.20), where the upper limit of the integral is a series expansion,[2] were solved in Appendix B of Ref. [24] by using contour integral approach in the complex-plane. They found that these integrals are a sum of two terms. The first term corresponds to a finite integral with its upper limit being the zeroth (Newtonian) order term of the expansion of $u_{(\max)}$ and the second term corresponds to the infinitely small or infinitely large contributions proportional to the PN corrections of the function $u_{(\max)}$. However, as the scattering angle is finite, the divergent terms entering the integral must cancel exactly with the divergent terms of Hadamard regularisation. Therefore, following Ref. [24], the correct value of a PN expanded real integral of Eq. (4.20) is obtained by first using the Newtonian limit of $u_{(\max)}$ as the upper limit of the integral and then taking the Hadamard partie finie (Pf) [25] of the integrals. The upper limit of the integral, $u_{(\max)}$, is the positive root closest to zero corresponding to the circular motion of Eq. (4.22), i.e.

$$\hat{\mathscr{E}}_{\text{eff}}^2 = A(u)\left(1 + j^2u^2\right), \tag{4.25}$$

and at Newtonian level, it reads

$$u_{(\max)} = \frac{1 + \sqrt{1 + 2\hat{E}j^2}}{j^2} + \mathscr{O}\left(\frac{1}{c^2}\right). \tag{4.26}$$

The integrals of Eq. (4.20) for ST theories can all be explicitly computed using the standard techniques (see Ref. [26]) except for one logarithmic integral arising at 3PN order which we compute in large j-expansion. To simplify the expressions of the scattering angle, we introduce an auxiliary variable[3]

$$\alpha = \frac{1}{\sqrt{\hat{E}j^2}}, \tag{4.27}$$

---

[2] In Ref. [24], they considered a more general integral where both the upper limit and the lower limit are series expansions.

[3] Note that the parameter α defined here is different from the ST parameters α_J, $\alpha(\varphi)$ and α_0.

and a function

$$B(\alpha) = \arctan(\alpha) + \frac{\pi}{2}. \tag{4.28}$$

The scattering angle for the local contribution $\frac{1}{2}\chi_{\text{loc}}$ up to 3PN order can be decomposed as a sum of contributions from each PN order, i.e.

$$\frac{1}{2}\chi_{\text{loc}}^{(\text{N})} = B(\alpha) - \frac{\pi}{2},$$

$$\frac{1}{2}\chi_{\text{loc}}^{(\text{1PN})} = \frac{1}{j^2}\left[C^{\text{B}}_{(1PN)} B(\alpha) + C^{0}_{(\text{1PN})}\right], \tag{4.29}$$

$$\frac{1}{2}\chi_{\text{loc}}^{(\text{2PN})} = \frac{1}{j^4}\left[C^{\text{B}}_{(\text{2PN})} B(\alpha) + C^{0}_{(\text{2PN})}\right], \tag{4.30}$$

$$\frac{1}{2}\chi_{\text{loc}}^{(\text{3PN})} = \frac{1}{j^6}\left[C^{\text{B}}_{(\text{3PN})} B(\alpha) + C^{0}_{(\text{3PN})}\right] + I_\chi, \tag{4.31}$$

where,

$$C^{\text{B}}_{(\text{1PN})} = \frac{1}{2}\left\{-a_{2,\text{ST}} + b_{1,\text{ST}}\right\} + 3,$$

$$C^{0}_{(\text{1PN})} = \frac{1}{2\alpha\,(1+\alpha^2)}\left\{-\alpha^2 a_{2,\text{ST}} + \alpha^2 b_{1,\text{ST}} + b_{1,\text{ST}}\right\} + \frac{3\alpha^2+2}{\alpha\,(1+\alpha^2)},$$

$$\begin{aligned} C^{\text{B}}_{(\text{2PN})} = \frac{1}{4\alpha^2}\Big[&105\alpha^2 - 3\alpha^2 a_{2,\text{ST}} b_{1,\text{ST}} - 42\alpha^2 a_{2,\text{ST}} + \frac{3}{2}\alpha^2 (a_{2,\text{ST}})^2 - 6\alpha^2 a_{3,\text{ST}} - 2a_{2,\text{ST}} \\ &+ 9\alpha^2 b_{1,\text{ST}} + 3\alpha^2 b_{2,\text{ST}} - \frac{3}{4}\alpha^2 b_{1,\text{ST}}^2 + b_{1,\text{ST}} + b_{2,\text{ST}} - \frac{b_{1,\text{ST}}^2}{4} + 15 + \left(-30\alpha^2 - 6\right)\nu\Big], \end{aligned}$$

$$\begin{aligned} C^{0}_{(\text{2PN})} = \frac{1}{4\alpha^2(1+\alpha^2)^2}\Big[&81\alpha + 190\alpha^3 + 105\alpha^5 - 3\alpha^5 a_{2,\text{ST}} b_{1,\text{ST}} - 5\alpha^3 a_{2,\text{ST}} b_{1,\text{ST}} - 2\alpha a_{2,\text{ST}} b_{1,\text{ST}} \\ &- 42\alpha^5 a_{2,\text{ST}} + \frac{3}{2}\alpha^5 a_{2,\text{ST}}^2 - 6\alpha^5 a_{3,\text{ST}} - 72\alpha^3 a_{2,\text{ST}} + \frac{5}{2}\alpha^3 a_{2,\text{ST}}^2 - 10\alpha^3 a_{3,\text{ST}} - 26\alpha a_{2,\text{ST}} \\ &- 4\alpha a_{3,\text{ST}} + 9\alpha^5 b_{1,\text{ST}} + 3\alpha^5 b_{2,\text{ST}} - \frac{3}{4}\alpha^5 b_{1,\text{ST}}^2 + 16\alpha^3 b_{1,\text{ST}} + 6\alpha^3 b_{2,\text{ST}} - \frac{3}{2}\alpha^3 b_{1,\text{ST}}^2 + 7\alpha b_{1,\text{ST}} \\ &+ 3\alpha b_{2,\text{ST}} - \frac{3}{4}\alpha b_{1,\text{ST}}^2 + \left(-30\alpha^5 - 56\alpha^3 - 26\alpha\right)\nu\Big], \end{aligned}$$

$$\begin{aligned} C^{\text{B}}_{(\text{3PN})} = &\frac{1}{\alpha^4}\left[\frac{9\nu^2}{8} - \frac{3\nu}{2} - \frac{3q_{3,\text{ST}}}{16}\right] + \frac{1}{\alpha^2}\Big[\frac{3}{32}a_{2,\text{ST}} b_{1,\text{ST}}^2 - \frac{21 a_{2,\text{ST}} b_{1,\text{ST}}}{8} - \frac{3a_{2,\text{ST}} b_{2,\text{ST}}}{8} + \frac{9\nu a_{2,\text{ST}}}{4} \\ &+ \frac{3a_{2,\text{ST}}^2}{2} - \frac{273 a_{2,\text{ST}}}{8} - \frac{3a_{3,\text{ST}} b_{1,\text{ST}}}{8} - \frac{27 a_{3,\text{ST}}}{4} - \frac{3a_{4,\text{ST}}^{\text{loc}}}{4} - \frac{3b_{1,\text{ST}} b_{2,\text{ST}}}{8} + \frac{3\nu b_{1,\text{ST}}}{2} + \frac{3b_{1,\text{ST}}^3}{32} \\ &- \frac{3b_{1,\text{ST}}^2}{16} + \frac{45 b_{1,\text{ST}}}{8} + \frac{9b_{2,\text{ST}}}{4} + \frac{3b_{3,\text{ST}}^{\text{loc}}}{4} + \frac{45\nu^2}{4} + \frac{123\pi^2\nu}{128} - 109\nu - \frac{9q_{3,\text{ST}}^{\text{loc}}}{8} + \frac{315}{4}\Big] \\ &+ \frac{15 a_{2,\text{ST}} a_{3,\text{ST}}}{4} + \frac{15}{16}a_{2,\text{ST}}^2 b_{1,\text{ST}} + \frac{15}{32}a_{2,\text{ST}} b_{1,\text{ST}}^2 - \frac{105 a_{2,\text{ST}} b_{1,\text{ST}}}{8} - \frac{15 a_{2,\text{ST}} b_{2,\text{ST}}}{8} + \frac{75\nu a_{2,\text{ST}}}{4} \\ &- \frac{1}{16}5a_{2,\text{ST}}^3 + \frac{195 a_{2,\text{ST}}^2}{8} - \frac{1485 a_{2,\text{ST}}}{8} - \frac{15 a_{3,\text{ST}} b_{1,\text{ST}}}{8} - \frac{135 a_{3,\text{ST}}}{4} - \frac{15 a_{4,\text{ST}}^{\text{loc}}}{4} - \frac{5b_{1,\text{ST}} b_{2,\text{ST}}}{8} + \frac{5b_{1,\text{ST}}^3}{32} \\ &+ \frac{175 b_{1,\text{ST}}}{8} - \frac{15 b_{1,\text{ST}}^2}{16} + \frac{25 b_{2,\text{ST}}}{4} + \frac{5b_{3,\text{ST}}^{\text{loc}}}{4} + \frac{105\nu^2}{8} + \frac{615\pi^2\nu}{128} - \frac{625\nu}{2} - \frac{15 q_{3,\text{ST}}^{\text{loc}}}{16} + \frac{1155}{4}, \end{aligned}$$

$$\begin{aligned} C^{0}_{(\text{3PN})} = \frac{1}{(1+\alpha^2)^3}\Big\{\alpha^5\Big[&\frac{15 a_{2,\text{ST}} a_{3,\text{ST}}}{4} + \frac{15}{16}a_{2,\text{ST}}^2 b_{1,\text{ST}} + \frac{15}{32}a_{2,\text{ST}} b_{1,\text{ST}}^2 - \frac{105 a_{2,\text{ST}} b_{1,\text{ST}}}{8} - \frac{15 a_{2,\text{ST}} b_{2,\text{ST}}}{8} \\ &+ \nu\left(\frac{75 a_{2,\text{ST}}}{4} + \frac{615\pi^2}{128} - \frac{625}{2}\right) - \frac{1}{16}5a_{2,\text{ST}}^3 + \frac{195 a_{2,\text{ST}}^2}{8} - \frac{1485 a_{2,\text{ST}}}{8} - \frac{15 a_{3,\text{ST}} b_{1,\text{ST}}}{8} - \frac{135 a_{3,\text{ST}}}{4} \\ &- \frac{15 a_{4,\text{ST}}^{\text{loc}}}{4} - \frac{5b_{1,\text{ST}} b_{2,\text{ST}}}{8} + \frac{5b_{1,\text{ST}}^3}{32} + \frac{175 b_{1,\text{ST}}}{8} - \frac{15 b_{1,\text{ST}}^2}{16} + \frac{25 b_{2,\text{ST}}}{4} + \frac{5b_{3,\text{ST}}^{\text{loc}}}{4} + \frac{105\nu^2}{8} - \frac{15 q_{3,\text{ST}}^{\text{loc}}}{16} \end{aligned}$$

$$
\begin{aligned}
&+\frac{1155}{4}\Bigg]+\alpha^3\Bigg[10a_{2,\mathrm{ST}}a_{3,\mathrm{ST}}+\nu\left(\frac{209a_{2,\mathrm{ST}}}{4}+\frac{3b_{1,\mathrm{ST}}}{2}+\frac{1763\pi^2}{128}-\frac{2827}{3}\right)+\frac{5}{2}a_{2,\mathrm{ST}}^2b_{1,\mathrm{ST}}\\
&+\frac{43}{32}a_{2,\mathrm{ST}}b_{1,\mathrm{ST}}^2-\frac{301a_{2,\mathrm{ST}}b_{1,\mathrm{ST}}}{8}-\frac{43a_{2,\mathrm{ST}}b_{2,\mathrm{ST}}}{8}-\frac{1}{6}5a_{2,\mathrm{ST}}^3+\frac{133a_{2,\mathrm{ST}}^2}{2}-\frac{4233a_{2,\mathrm{ST}}}{8}-\frac{43a_{3,\mathrm{ST}}b_{1,\mathrm{ST}}}{8}\\
&-\frac{387a_{3,\mathrm{ST}}}{4}-\frac{43a_{4,\mathrm{ST}}^{\mathrm{loc}}}{4}-\frac{49b_{1,\mathrm{ST}}b_{2,\mathrm{ST}}}{24}+\frac{49b_{1,\mathrm{ST}}^3}{96}+\frac{1535b_{1,\mathrm{ST}}}{24}-\frac{43b_{1,\mathrm{ST}}^2}{16}+\frac{227b_{2,\mathrm{ST}}}{12}+\frac{49b_{3,\mathrm{ST}}^{\mathrm{loc}}}{12}\\
&+\frac{185\nu^2}{4}-\frac{29q_{3,\mathrm{ST}}^{\mathrm{loc}}}{8}+\frac{3395}{4}\Bigg]+\alpha\Bigg[\frac{33a_{2,\mathrm{ST}}a_{3,\mathrm{ST}}}{4}+\nu\left(\frac{189a_{2,\mathrm{ST}}}{4}+4b_{1,\mathrm{ST}}+\frac{1681\pi^2}{128}-\frac{2939}{3}\right)\\
&+\frac{33}{16}a_{2,\mathrm{ST}}^2b_{1,\mathrm{ST}}+\frac{41}{32}a_{2,\mathrm{ST}}b_{1,\mathrm{ST}}^2-\frac{287a_{2,\mathrm{ST}}b_{1,\mathrm{ST}}}{8}-\frac{41a_{2,\mathrm{ST}}b_{2,\mathrm{ST}}}{8}-\frac{1}{16}11a_{2,\mathrm{ST}}^3+\frac{461a_{2,\mathrm{ST}}^2}{8}\\
&-\frac{3995a_{2,\mathrm{ST}}}{8}-\frac{41a_{3,\mathrm{ST}}b_{1,\mathrm{ST}}}{8}-\frac{369a_{3,\mathrm{ST}}}{4}-\frac{41a_{4,\mathrm{ST}}^{\mathrm{loc}}}{4}-\frac{19b_{1,\mathrm{ST}}b_{2,\mathrm{ST}}}{8}+\frac{19b_{1,\mathrm{ST}}^3}{32}+\frac{505b_{1,\mathrm{ST}}}{8}\\
&-\frac{41b_{1,\mathrm{ST}}^2}{16}+\frac{79b_{2,\mathrm{ST}}}{4}+\frac{19b_{3,\mathrm{ST}}^{\mathrm{loc}}}{4}+60\nu^2-\frac{21q_{3,\mathrm{ST}}^{\mathrm{loc}}}{4}+\frac{3381}{4}\Bigg]+\frac{1}{\alpha}\Big[2a_{2,\mathrm{ST}}a_{3,\mathrm{ST}}\\
&+\nu\left(\frac{55a_{2,\mathrm{ST}}}{4}+\frac{7b_{1,\mathrm{ST}}}{2}+\frac{533\pi^2}{128}-\frac{1153}{3}\right)+\frac{13}{32}a_{2,\mathrm{ST}}b_{1,\mathrm{ST}}^2+\frac{1}{2}a_{2,\mathrm{ST}}^2b_{1,\mathrm{ST}}-\frac{95a_{2,\mathrm{ST}}b_{1,\mathrm{ST}}}{8}\\
&-\frac{13a_{2,\mathrm{ST}}b_{2,\mathrm{ST}}}{8}+\frac{29a_{2,\mathrm{ST}}^2}{2}-\frac{1279a_{2,\mathrm{ST}}}{8}-\frac{13a_{3,\mathrm{ST}}b_{1,\mathrm{ST}}}{8}-\frac{121a_{3,\mathrm{ST}}}{4}-\frac{13a_{4,\mathrm{ST}}^{\mathrm{loc}}}{4}-\frac{9b_{1,\mathrm{ST}}b_{2,\mathrm{ST}}}{8}\\
&+\frac{9b_{1,\mathrm{ST}}^3}{32}-\frac{13b_{1,\mathrm{ST}}^2}{16}+\frac{179b_{1,\mathrm{ST}}}{8}+\frac{31b_{2,\mathrm{ST}}}{4}+\frac{9b_{3,\mathrm{ST}}^{\mathrm{loc}}}{4}+\frac{135\nu^2}{4}-\frac{27q_{3,\mathrm{ST}}^{\mathrm{loc}}}{8}+\frac{1221}{4}\Bigg]\\
&+\frac{1}{\alpha^3}\Bigg[-\frac{a_{2,\mathrm{ST}}b_{1,\mathrm{ST}}}{2}-6a_{2,\mathrm{ST}}-a_{3,\mathrm{ST}}-\frac{b_{1,\mathrm{ST}}b_{2,\mathrm{ST}}}{6}+\nu\left(b_{1,\mathrm{ST}}-\frac{69}{2}\right)+\frac{b_{1,\mathrm{ST}}^3}{24}+\frac{4b_{1,\mathrm{ST}}}{3}\\
&+\frac{2b_{2,\mathrm{ST}}}{3}+\frac{b_{3,\mathrm{ST}}^{\mathrm{loc}}}{3}+\frac{55\nu^2}{8}-\frac{13q_{3,\mathrm{ST}}^{\mathrm{loc}}}{16}+\frac{64}{3}\Bigg]\Bigg\}.
\end{aligned}
$$

Here, for simplicity, we do not substitute the values of the ST corrections $a_{i,\mathrm{ST}}$, $b_{i,\mathrm{ST}}$ and $q_{i,\mathrm{ST}}$. The explicit expressions of the corrections have been derived in Refs. [16–20].

Finally, the last contribution, I_χ, to the 3PN scattering angle given in Eq. (4.31) is determined by inserting 3PN order $\hat{p}_r$ from Eq. (4.23) into Eq. (4.20) as the following integral,

$$I_\chi = -\frac{j a_{4,\mathrm{ST}}^{\log}}{2}\mathrm{Pf}\int_0^{u_{(max)}} \frac{u^4 \ln(u)}{\left(2\hat{E}+2u-j^2u^2\right)^{3/2}}du. \tag{4.32}$$

The above integral cannot be solved explicitly using the standard techniques of integration therefore we simplify the integral by using suitable integration by parts according to

$$I_\chi = a_{4,\log}^{\mathrm{ST}}\frac{(15\alpha^4+18\alpha^2+3)}{16j^6\alpha^2(\alpha^2+1)}B(\alpha)+a_{4,\mathrm{ST}}^{\log}\frac{(15\alpha^2+13)}{16(\alpha^2+1)\alpha j^6}+\mathscr{I}_\chi, \tag{4.33}$$

where the last term is now a *convergent* integral given by

$$\mathscr{I}_\chi = \frac{2 j a_{4,\mathrm{ST}}^{\log}}{(1+2j^2\hat{E})}\int_0^{u_{(max)}} \frac{u^3(uj^2-1)\ln(u)}{\sqrt{2\hat{E}+2u-j^2u^2}}du. \tag{4.34}$$

The integral of Eq. (4.34) cannot be expressed in terms of the elementary functions. After a suitable change of variables,

$$u = \frac{\sqrt{2\hat{E}}}{j} x' \quad ; \quad \varepsilon \equiv 2\alpha = \frac{2}{p_\infty j}, \tag{4.35}$$

the integral can be computed in the large j-expansion (small ε-expansion) at fixed p_∞. We follow the approach of Ref. [9] to compute the j-expansion of the integral. Here, we display the first three contributions to the integral in the j-expansion,

$$\mathscr{I}_\chi = \frac{a_{4,\mathrm{ST}}^{\log} \hat{E}}{j^4} \mathscr{I}_4 + \frac{a_{4,\mathrm{ST}}^{\log} \hat{E}^{1/2}}{j^5} \mathscr{I}_5 + \frac{a_{4,\mathrm{ST}}^{\log}}{j^6} \mathscr{I}_6 + \mathscr{O}\left(\frac{\hat{E}^{-1/2}}{j^7}\right), \tag{4.36}$$

where

$$\mathscr{I}_4 = \frac{\pi}{16}\left[7 + 6\ln\left(\frac{\hat{E}}{2j^2}\right)\right],$$

$$\mathscr{I}_5 = \sqrt{2}\left[-2 + 8\ln(2) + 2\ln\left(\frac{\hat{E}}{2j^2}\right)\right],$$

$$\mathscr{I}_6 = \frac{\pi}{32}\left[77 + 30\ln\left(\frac{\hat{E}}{2j^2}\right)\right].$$

The higher-order contributions can be computed following the same approach.

4.3.2 Final Expression of the Local Part of the 3PN Scattering Angle in the Large j-Expansion

The result presented in Eq. (4.31) for the scattering angle at the 3PN-order is unexpanded in j except for the integral I_χ of Eq. (4.32). To compute this integral, we express it as a simpler integral of Eq. (4.33). Then at the end of the last subsection, we have computed the large j-expansion of this remaining part, $\mathscr{I}_\chi$, of the integral I_χ.

Let us now present the large j-expansion $\chi_{\mathrm{loc}}^{(3\mathrm{PN})}/2$ given by Eq. (4.31). Note that the auxiliary variable α entering the exactly known part of the scattering angle at 3PN, $\chi_{\mathrm{loc}}^{(3\mathrm{PN})}/2$ is also dependent on j [see, Eq. (4.27)]. As the contributions to $\mathscr{I}_\chi$, and hence I_χ given by Eq. (4.33) start at $1/j^4$ order in their large j-expansion, we here present the large j-expansion of $\chi_{\mathrm{loc}}^{(3\mathrm{PN})}/2$ up to $1/j^4$,

$$\frac{\chi_{3\mathrm{PN}}}{2} - \mathscr{I}_\chi = \frac{p_\infty^4}{j^2}\left[-\frac{3}{32}\pi q_{3,\mathrm{ST}}^{\mathrm{loc}} + \frac{9}{16}\pi\nu^2 - \frac{3}{4}\pi\nu\right] + \frac{p_\infty^3}{j^3}\left[-\frac{a_{2,\mathrm{ST}} b_{1,\mathrm{ST}}}{2} - 6a_{2,\mathrm{ST}} - a_{3,\mathrm{ST}} - \frac{b_{1,\mathrm{ST}} b_{2,\mathrm{ST}}}{6}\right.$$
$$\left. + \nu b_{1,\mathrm{ST}} + \frac{b_{1,\mathrm{ST}}^3}{24} + \frac{4b_{1,\mathrm{ST}}}{3} + \frac{2b_{2,\mathrm{ST}}}{3} + \frac{b_{3,\mathrm{ST}}^{\mathrm{loc}}}{3} + 8\nu^2 - 36\nu - q_{3,\mathrm{ST}}^{\mathrm{loc}} + \frac{64}{3}\right] + \frac{p_\infty^2}{j^4}\left[\frac{3}{64}\pi a_{2,\mathrm{ST}} b_{1,\mathrm{ST}}^2\right.$$

$$
\begin{aligned}
&-\frac{21}{16}\pi a_{2,\mathrm{ST}}b_{1,\mathrm{ST}}-\frac{3}{16}\pi a_{2,\mathrm{ST}}b_{2,\mathrm{ST}}+\frac{9}{8}\pi\nu a_{2,\mathrm{ST}}+\frac{3\pi a_{2,\mathrm{ST}}^2}{4}-\frac{273\pi a_{2,\mathrm{ST}}}{16}-\frac{3}{16}\pi a_{3,\mathrm{ST}}b_{1,\mathrm{ST}}\\
&-\frac{27\pi a_{3,\mathrm{ST}}}{8}-\frac{3\pi a_{4,\mathrm{ST}}^{\mathrm{loc}}}{8}+\frac{3\pi a_{4,\mathrm{ST}}^{\mathrm{log}}}{32}-\frac{3}{16}\pi b_{1,\mathrm{ST}}b_{2,\mathrm{ST}}+\frac{3}{4}\pi\nu b_{1,\mathrm{ST}}+\frac{3\pi b_{1,\mathrm{ST}}^3}{64}-\frac{3\pi b_{1,\mathrm{ST}}^2}{32}\\
&+\frac{45\pi b_{1,\mathrm{ST}}}{16}+\frac{9\pi b_{2,\mathrm{ST}}}{8}+\frac{3\pi b_{3,\mathrm{ST}}^{\mathrm{loc}}}{8}+\frac{45\pi\nu^2}{8}+\frac{123\pi^3\nu}{256}-\frac{109\pi\nu}{2}-\frac{9\pi q_{3,\mathrm{ST}}^{\mathrm{loc}}}{16}+\frac{315\pi}{8}\\
&+\frac{\pi a_{4,\mathrm{ST}}^{\mathrm{log}}}{32}\left\{7+6\ln\left(\frac{\hat{E}}{2j^2}\right)\right\}\Bigg]+\mathcal{O}\left(\frac{1}{j^5}\right).
\end{aligned}
\tag{4.37}
$$

4.4 Scalar-Tensor Scattering Angle: Non-local Contributions

In this section, we compute the leading order nonlocal contributions to the scattering angle using the order-reduction approach of Ref. [15] for bound orbits. This approach has recently been used to derive the nonlocal contributions to the EOB metric potentials for bound orbits in ST theories [19, 20]. We will use this approach for hyperbolic orbits in ST theories, following Refs. [9, 27].

Since the tail contribution to the Hamiltonian starts at 3PN order in ST theory, one can compute the leading order contribution to the scattering angle χ_{nonloc} by considering the Hamiltonian

$$
H - H_{\mathrm{N}} + H^{\mathrm{tail}}, \tag{4.38}
$$

where H_{N} is the Newtonian-order Hamiltonian and H^{tail} is the leading order tail contribution [14], as only the Newtonian order radial momentum is required for computing the 3PN leading order tail contribution to the scattering angle. The Newtonian order contribution to the Hamiltonian in terms of dimensionless variables given in Eq. (4.3) reads,

$$
\frac{H_{\mathrm{N}}}{\mu} = \frac{1}{2}\left(\hat{p}_r^2 + \frac{j^2}{\hat{r}^2}\right) - \frac{1}{\hat{r}}. \tag{4.39}
$$

For simplicity we work in units where $M = 1$ such that $\mu = \nu$ in this section.

From Sect. 4.3, we recall the formula for the scattering angle derived using the general Hamilton-Jacobi approach,

$$
\chi(\hat{E}, j) = -\frac{\partial}{\partial j}\int \hat{p}_r(\hat{E}, j, \hat{r})d\hat{r}, \tag{4.40}
$$

where the radial momentum function, $\hat{p}_r(\hat{E}, j, r)$, is first computed by solving the energy conservation law,

$$
\hat{E} = \frac{H(\hat{r}, \hat{p}_r, j)}{\nu} = \frac{1}{2}\left(\hat{p}_r^2 + \frac{j^2}{\hat{r}^2}\right) - \frac{1}{\hat{r}} + \frac{H^{\mathrm{tail}}}{\nu} \tag{4.41}
$$

for $\hat{p}_r^2(\hat{E}, j, r)$. At leading order in the tail, the solution of the equation in terms of $\hat{p}_r$ is

$$\hat{p}_r = \hat{p}_r^0 - \frac{1}{\hat{p}_r^0}\frac{H^{\text{tail}}(\hat{r}, \hat{p}_r^0, j)}{\nu}, \tag{4.42}$$

where $\hat{p}_r^0$ is the Newtonian contribution (see, Eq. (4.24)). Inserting the solution into Eq. (4.40), we obtain

$$\chi(\hat{E}, j) = -\frac{\partial}{\partial j}\int \hat{p}_r^0(\hat{E}, j, \hat{r})dr + \frac{\partial}{\partial j}\int \frac{dr}{\hat{p}_r^0}\frac{H^{\text{tail}}(\hat{r}, \hat{p}_r, j)}{\nu}, \tag{4.43}$$

where the first term is the Newtonian order contribution to the scattering angle derived in Eq. (4.29). Then the nonlocal contribution to the scattering angle reads

$$\chi_{\text{nonloc}} = \frac{1}{\nu}\frac{\partial}{\partial j}W^{\text{tail}}(\hat{E}, j), \tag{4.44}$$

where

$$W^{\text{tail}} = \int \frac{d\hat{r}}{\hat{p}_r^0}H^{\text{tail}}(\hat{r}, \hat{p}_r, j) = \left[\!\!\left[\int dt H^{\text{tail}}\right]\!\!\right]. \tag{4.45}$$

In the second expression we have used the property that the time localised Hamiltonian is simply obtained by using the solutions of the Newtonian-level Hamilton's equation for the phase space variables (see, Ref. [9] for the hyperbolic case and Ref. [15] for the elliptic case), and that $d\hat{r}/\hat{p}_r^0 = dt$ is along the Newtonian Hamiltonian flow.

The leading order tail contribution to the Hamiltonian in ST theories in the Jordan-Frame conventions reads [14]

$$H_{\text{LO}}^{\text{tail}} = -\frac{2}{3}(3 + 2\omega_0)\text{Pf}_{2s_1/c}\int_{-\infty}^{\infty}\frac{d\tau}{|\tau|}I_{s,i}^{(2)}(t)I_{s,i}^{(2)}(t+\tau), \tag{4.46}$$

where s_1 is the Hadamard partie finie scale, and $I_{s,i}^{(2)}$ is the second time-derivative of the scalar dipole moment, $I_{s,i}$. In the centre-of-mass (COM) frame, the scalar dipole moment in the DEF convention is

$$I_{s,i} = \nu\alpha_0(\alpha_B - \alpha_A)\hat{x}^i, \tag{4.47}$$

where α_A, α_B are given in Table 2.1 and α_0 is the asymptotic value of the parameter $\alpha(\varphi)$ at $\varphi = \varphi_0$. The parameter $\alpha(\varphi)$ is defined using the coupling function $\mathscr{A}$ (see Eq. (2.3)) as

$$\alpha(\varphi) = \frac{\partial \ln \mathscr{A}}{\partial \varphi}, \tag{4.48}$$

and measures the coupling between the matter and the scalar fields in the ST equations of motion. Here, $\hat{x}^i (= (Z_A - Z_B)^i/(G_{AB}M))$ is the dimensionless relative separation vector and $Z_{A,B}$ indicate the positions of the two bodies.

Thus the leading order potential $W^{\rm tail}$ is,

$$W^{\rm tail} = \int dt \; H^{\rm tail}_{\rm LO}, \tag{4.49}$$

where $H^{\rm tail}_{\rm LO}$ is given in Eq. (4.46).

In Refs. [19, 20], using Kepler's equations for elliptic orbits, it has been shown that the scalar dipole moment is a periodic function in the action-angle variables and, hence can be decomposed into a Fourier series. Here, we are considering hyperbolic motions, therefore the Cartesian coordinates are parameterised as

$$x(t) = -\, a(\cosh\bar{u} - e), \tag{4.50}$$

$$y(t) = -\, a\sqrt{e^2 - 1}\, \sinh\bar{u}, \tag{4.51}$$

and the hyperbolic Kepler equation is

$$\bar{n}t = e\, \sinh\bar{u} - \bar{u}, \tag{4.52}$$

where e is the eccentricity, a is the semi-major axis, $\bar{u}$ is the eccentric anomaly, and

$$\bar{n} = \frac{1}{\bar{a}^{3/2}}; \quad \bar{a} = -a = \frac{1}{2\hat{E}}. \tag{4.53}$$

Similar to the Fourier series expansion of the scalar dipole moment for elliptic orbits in Refs. [16, 20], the scalar dipole moment for hyperbolic motion can also be decomposed into a Fourier series, i.e.

$$\begin{aligned} I_{\rm s,i}(t) &= \int \frac{d\omega}{2\pi} \tilde{I}_{s,i}(\omega) {\rm e}^{-i\omega t}, \\ \tilde{I}_{s,i}(\omega) &= \int dt\; I_{s,i}(t) {\rm e}^{i\omega t}, \end{aligned} \tag{4.54}$$

where $\tilde{I}_{s,i}(\omega)$ is the Fourier-transform of the scalar dipole moment.

Inserting the Fourier transformation of the scalar dipole moment (and $\tau = G_{AB}M(t' - t))$[4] in Eq. (4.49) and converting to the DEF convention yields,

$$W_{\rm tail} = -\frac{1}{(1+\alpha_A\alpha_B)^2}\frac{2}{3\alpha_0^2}\int dt\; {\rm Pf}_{2\hat{s}_1/c}\int \frac{dt'}{|t-t'|}\int \frac{d\omega\, d\omega'}{2\pi\; 2\pi}\omega^2\omega'^2 \tilde{I}_{s,i}(\omega)\tilde{I}_{s,i}(\omega'){\rm e}^{-i\omega t}{\rm e}^{-i\omega' t'}$$

---

[4] Here, the dimensionless variable, $\hat{\tau} = \frac{\tau}{G_{AB}}$ and $t = \frac{T}{G_{AB}M}$.

$$
\begin{aligned}
&= -\frac{2}{3\alpha_0^2(1+\alpha_A\alpha_B)^2}\mathrm{Pf}_{2\hat{s}_1/c}\int\frac{dt'}{|t-t'|}\int\frac{d\omega\, d\omega'}{2\pi\ 2\pi}\omega^2\omega'^2\tilde{I}_{s,i}(\omega)\tilde{I}_{s,i}(\omega')\mathrm{e}^{-i\omega'\hat{\tau}}\,2\pi\delta(\omega+\omega')\\
&= -\frac{2}{3\alpha_0^2(1+\alpha_A\alpha_B)^2}\mathrm{Pf}_{2\hat{s}_1/c}\int\frac{dt'}{|t-t'|}\int\frac{d\omega}{2\pi}\omega^4\tilde{I}_{s,i}(\omega)\tilde{I}_{s,i}(-\omega)\mathrm{e}^{i\omega\hat{\tau}}\\
&= -\frac{2}{3\alpha_0^2(1+\alpha_A\alpha_B)^2}\int\frac{d\omega}{2\pi}\omega^4\tilde{I}_{s,i}(\omega)\tilde{I}_{s,i}(-\omega)\mathrm{Pf}_{2\hat{s}_1/c}\int\frac{d\hat{\tau}}{|\hat{\tau}|}\mathrm{e}^{i\omega\hat{\tau}}.
\end{aligned}
\tag{4.55}
$$

The partie finie integral of the last term in the above equation is [see, Ref. [28]],

$$
\mathrm{Pf}_{2\hat{s}_1/c}\int_{-\infty}^{\infty}\frac{d\tau}{|\tau|}\mathrm{e}^{i\omega\tau} = -2\ln\left(\frac{2|\omega|\hat{s}_1\,\mathrm{e}^{\gamma_\mathrm{E}}}{c}\right), \tag{4.56}
$$

where γ_E is Euler's constant. Inserting Eq. (4.56) into Eq. (4.55) gives the Fourier-domain formula for the potential,

$$
W^{\mathrm{tail}} = \frac{8}{3\alpha_0^2(1+\alpha_A\alpha_B)^2}\int_0^{\infty}\frac{d\omega}{2\pi}\omega^4\tilde{I}_{s,i}(\omega)\tilde{I}^*_{s,i}(\omega)\ln\left(\frac{2|\omega|\hat{s}_1\,\mathrm{e}^{\gamma_\mathrm{E}}}{c}\right), \tag{4.57}
$$

where $\tilde{I}_{s,i}(-\omega) = \tilde{I}^*_{s,i}(\omega)$.

To compute the explicit expression of Eq. (4.57) for the potential W^{tail} in terms of $\hat{E}$ and j, we insert the Fourier transform of the scalar dipole moment. For this, we evaluate the Fourier transforms of (x, y) for hyperbolic orbits, i.e.

$$
\begin{aligned}
\tilde{x}(\omega) &= \int dt\ \mathrm{e}^{i\omega t}x(t),\\
\tilde{y}(\omega) &= \int dt\ \mathrm{e}^{i\omega t}y(t).
\end{aligned}
\tag{4.58}
$$

After inserting Eqs. (4.50)–(4.52) into Eq. (4.58) and using the definition of Hankel functions of the first kind $H^{(1)}_{p'}$ (see Eq. (9.1.25) of [29]),

$$
\int_{-\infty}^{\infty}\mathrm{e}^{q'\sinh\xi - p'\xi} = i\pi H^{(1)}_{p'}(q'), \tag{4.59}
$$

we find the Fourier transform of (x, y). The computation gives

$$
\tilde{x}(\omega) = \frac{\pi a}{\omega}\left(\frac{p'_\omega}{q'_\omega}H^{(1)}_{p'_\omega}(q'_\omega) - H^{(1)}_{p'_\omega+1}(q'_\omega)\right), \tag{4.60}
$$

$$
\tilde{y}(\omega) = -\frac{\pi a}{\omega e}\sqrt{e^2-1}\,H^{(1)}_{p}(q), \tag{4.61}
$$

where in our case

$$
q' = q'_\omega = ie\frac{\omega}{\bar{n}};\quad p' = p'_\omega = \frac{q'_\omega}{e}. \tag{4.62}
$$

We then consider the Fourier transform of (x, y) in the large-j limit which is equivalent to the large-e limit since $e = \sqrt{1 + 2\hat{E} j^2}$. The large-$e$ limit of Eqs. (4.60)–(4.61) yields,

$$\tilde{x}(\omega) = -\frac{\pi a}{\omega} H_1^{(1)}(i g_\omega), \tag{4.63}$$

$$\tilde{y}(\omega) = -\frac{\pi a}{\omega} H_0^{(1)}(i g_\omega), \tag{4.64}$$

where $q'_\omega = i g_\omega$ and $g_\omega = e\frac{\omega}{\bar{n}}$. The Hankel functions evaluated at purely imaginary arguments are related to the modified Bessel functions K_ν by (see Eq. (9.6.4) of [29])

$$K_0(x') = i\frac{\pi}{2} H_0^{(1)}(i x')\ ; \quad K_1(x') = -\frac{\pi}{2} H_1^{(1)}(i x'). \tag{4.65}$$

Finally, inserting Eqs. (4.63)–(4.65) in Eq. (4.57) and then taking the j-derivative of the potential W^{tail}, the explicit expression for the scattering angle in the large-e limit yields,

$$\chi_{\text{3PN}}^{\text{tail}} = -\frac{2\pi\nu}{3}\frac{p_\infty^2}{j^4}\left[2\delta_+ + \frac{\bar{\gamma}_{AB}(\bar{\gamma}_{AB}+2)}{2}\right]\left\{7 + 3\ln\left(\frac{p_\infty^2 \hat{s}_1}{4j}\right)\right\}, \tag{4.66}$$

where we recall that $\hat{s}_1 = s_1/(G_{AB}M)$ is the dimensionless regularisation scale defining the nearzone-farzone separation, $\delta_+ = \frac{\delta_A+\delta_B}{2}$, and $\bar{\gamma}_{AB}$ is defined in Table 2.1.

4.5 Summing the Local and Non-local Contributions to χ_{3PN} in the Large-J Expansion

In Sect. 4.3, we first computed the local scattering angle up to 3PN order and then in Sect. 4.4 we separately computed the nonlocal contributions at 3PN order. The results at 1PN and 2PN levels were given in a fully explicit and exact form. However, the results at 3PN order were obtained in the large-j expansion for both the local contribution (due to the logarithmic term $\mathscr{I}_\chi$) and the nonlocal contribution. On combining the two separate 3PN order contributions to the scattering angle at 3PN order, we find

$$\begin{aligned}
\frac{\chi(\hat{E}, j)^{(3\text{PN})}}{2} &= \frac{p_\infty^4}{j^2}\pi\left\{\nu\left[-\frac{3}{4} - \frac{15}{64}\bar{\gamma}_{\text{AB}}^2 - \frac{13}{16}\bar{\gamma}_{\text{AB}} - \frac{1}{16}\langle\bar{\beta}\rangle + \frac{1}{16}\langle\delta\rangle\right] + \nu^2\left[\frac{3}{8}\bar{\gamma}_{\text{AB}} - \frac{3}{16}\langle\bar{\beta}\rangle + \frac{9}{16}\right]\right\} \\
&+ \frac{p_\infty^3}{j^3}\left\{\frac{64}{3} - \frac{4}{3}\left[\langle\delta\rangle(\bar{\gamma}_{\text{AB}}+2) - 2\bar{\gamma}_{\text{AB}}\left(\bar{\gamma}_{\text{AB}}^2 + 6\bar{\gamma}_{\text{AB}} + 12\right)\right] + \frac{\nu}{6}\left[-216 + 4\langle\bar{\beta}\rangle(8\bar{\gamma}_{\text{AB}} - 3)\right.\right. \\
&+ 8\bar{\gamma}_{\text{AB}}\delta_+ - 6\bar{\gamma}_{\text{AB}}^3 - 93\bar{\gamma}_{\text{AB}}^2 - 262\bar{\gamma}_{\text{AB}} - 4X_{\text{AB}}\bar{\gamma}_{\text{AB}}\beta_- + 8\langle\delta\rangle - 57X_{\text{AB}}\beta_- + 2X_{\text{AB}}\varepsilon_- \\
&\left.\left. + 36\beta_+ + 4\delta_+ - 2\zeta - 2\varepsilon_+\right] + \frac{\nu^2}{3}\left[\bar{\gamma}_{\text{AB}}^2 + 22\bar{\gamma}_{\text{AB}} - 6\langle\bar{\beta}\rangle - 18\beta_+ + 4\delta_+ - 6\zeta + 2\varepsilon_+ + 24\right]\right\} \\
&+ \frac{p_\infty^2}{j^4}\pi\left\{\frac{315}{8} + \frac{1}{32}\left[4\langle\delta\rangle\left(-20\bar{\gamma}_{\text{AB}} + 12\beta_+^3 + 12\beta_-^2\beta_+ - 35\right) + \langle\bar{\beta}\rangle\left(-187\bar{\gamma}_{\text{AB}}^2 - 556\bar{\gamma}_{\text{AB}}\right.\right.\right.
\end{aligned}$$

$$
\begin{aligned}
&+12\langle\delta\rangle-412)+16\bar{\gamma}_{\mathrm{AB}}\langle\varepsilon\rangle+236\bar{\gamma}_{\mathrm{AB}}^{3}+1229\bar{\gamma}_{\mathrm{AB}}^{2}+2148\bar{\gamma}_{\mathrm{AB}}+48\langle\bar{\beta}\rangle^{2}-8\langle\kappa\rangle+24\langle\varepsilon\rangle\\
&+48\beta_{-}\left(\beta_{-}^{2}+\beta_{+}^{2}\right)(18\beta_{+}-\delta_{+})X_{\mathrm{AB}}-48\beta_{-}\delta_{-}(\beta_{-}^{2}+\beta_{+}^{2})\Big]+\nu\Bigg[\frac{123\pi^{2}}{256}-\frac{109}{2}\\
&+\bar{\gamma}_{\mathrm{AB}}\left(\frac{49\langle\bar{\beta}\rangle}{4}-\frac{3\langle\delta\rangle}{16}-\frac{3\beta_{-}X_{\mathrm{AB}}}{4}+\frac{39\beta_{+}}{4}+\frac{\delta_{+}}{2}-\frac{21\pi^{2}\delta_{+}}{256}+\frac{15\zeta}{4}-\varepsilon_{+}+\frac{225\pi^{2}}{512}-\frac{3353}{48}\right)\\
&+\left(\frac{\langle\bar{\beta}\rangle}{4}+\frac{15\pi^{2}}{256}-\frac{2425}{96}\right)\bar{\gamma}_{\mathrm{AB}}^{2}+\left(-\frac{71}{64}-\frac{21\pi^{2}}{1024}\right)\bar{\gamma}_{\mathrm{AB}}^{3}+\left(\delta_{+}+\frac{7}{4}-\frac{3}{4}\langle\bar{\beta}\rangle\right)\langle\bar{\beta}\rangle\\
&+\left(\frac{9\beta_{-}X_{\mathrm{AB}}}{2}+\frac{81}{4}\right)\beta_{+}+\frac{9}{4}(\beta_{-}^{2}+\beta_{+}^{2})-\left(\frac{17}{12}+\frac{21\pi^{2}}{128}\right)\delta_{+}+\frac{9}{8}\langle\delta\rangle+\left(-\frac{45\beta_{-}}{4}-\frac{\delta_{-}}{8}\right.\\
&\left.+\frac{3\varepsilon_{-}}{8}\right)X_{\mathrm{AB}}+\beta_{-}\delta_{-}+3\zeta+\frac{\kappa_{+}}{2}-\frac{15\varepsilon_{+}}{8}+\frac{\langle\kappa\rangle}{4}-\frac{3}{8}\langle\varepsilon\rangle+\left(2\delta_{+}+\frac{\bar{\gamma}_{\mathrm{AB}}(\bar{\gamma}_{\mathrm{AB}}+2)}{2}\right)\ln\left(\frac{2}{p_{\infty}}\right)\\
&+\frac{12\left(\beta_{-}^{2}-\beta_{+}^{2}\right)\langle\bar{\beta}\rangle}{\bar{\gamma}_{\mathrm{AB}}^{2}}-\frac{3\bar{\gamma}_{\mathrm{AB}}\left(11\left(\bar{\gamma}_{\mathrm{AB}}+2\right)^{2}-4\langle\delta\rangle\right)}{32\alpha_{\mathrm{AB}}\left(\bar{\gamma}_{\mathrm{AB}}+2\right)}+\frac{\beta_{+}\left(3\langle\varepsilon\rangle+4\delta_{-}X_{\mathrm{AB}}-8\delta_{+}\right)}{2\bar{\gamma}_{\mathrm{AB}}}\\
&+\frac{\beta_{-}\left(X_{\mathrm{AB}}\left(4\delta_{+}+3\varepsilon_{+}\right)-8\delta_{-}-3\varepsilon_{-}\right)}{2\bar{\gamma}_{\mathrm{AB}}}\Bigg]-\frac{3}{8}\nu^{2}\Big[-\bar{\gamma}_{\mathrm{AB}}^{2}-16\bar{\gamma}_{\mathrm{AB}}+3-4\beta_{-}^{2}+18\beta_{+}-4\delta_{+}\\
&+6\zeta-2\varepsilon_{+}-15\Big]\Big\}, \qquad (4.67)
\end{aligned}
$$

using the short-hand notations of Chaps. 2 and 3 namely,

$$X_{AB} \equiv X_A - X_B, \qquad (4.68)$$

$$\langle\bar{\beta}\rangle \equiv -X_{AB}\beta_- + \beta_+, \qquad (4.69)$$

$$\langle\bar{\kappa}\rangle \equiv -X_{AB}\kappa_- + \kappa_+, \qquad (4.70)$$

$$\langle\delta\rangle \equiv X_{AB}\delta_- + \delta_+, \qquad (4.71)$$

$$\langle\varepsilon\rangle \equiv -X_{AB}\varepsilon_- + \varepsilon_+, \qquad (4.72)$$

where $X_{\mathrm{A,B}} \equiv m^0_{A,B}/M$. Here, the subscript "±" denotes the symmetric and anti-symmetric parts of the ST parameters, e.g. $z_\pm = (z_A \pm z_B)/2$. Since the scattering angle is gauge-invariant, the arbitrary Hadamard Partie finie scale $\hat{s}_1$ cancels between the local contribution and the non-local contribution, as expected. This can be seen by using the explicit expressions of local and non-local contributions to the EOB metric potentials A, B, Q in Eqs. (4.37) and (4.66) following Ref. [20].

4.6 Conclusions

Building upon the results of Refs. [16–20] for the corrections in the EOB metric coefficients (A, B, Q_e) for massless scalar-tensor theory for the conservative part of the dynamics, we have determined the scattering angle for hyperboliclike orbits up to the 3PN order for both the local-in-time and the nonlocal-in-time parts of the dynamics. First, we have computed the scattering angle for the local part of the dynamics by (i) deriving the radial momentum as a function of u, orbital angular momentum and energy by iteratively solving the EOB energy conservation law; (ii) calculating the scattering angle using the standard techniques of Ref. [24] for solving divergent integrals arising in the PN-expansion of the radial momentum except the integral I_χ at the 3PN order; and (iii) computing the integral I_χ by using the appropriate integration by parts and expanding in large j the remaining integral after a change of variables [9]. We then computed the total contribution to the 3PN order scattering angle in the large-j expansion.

Then, we computed the nonlocal-in-time contribution by using the approach introduced in Ref. [15] for GR of order-reducing (time-localisation) the Hamiltonian in the small-eccentricity case for hyperboliclike encounters [9, 27]. Finally, we substituted the ST corrections of the metric potentials (A, B, Q_e) and summed both the local and nonlocal contributions in the large-j expansion at 3PN order. As the first test of our results, we checked that the scattering angle coincides with the scattering angle of GR (see Ref. [9] for GR results) in the GR limit as expected.

Another test we have performed on our results is to study the binary black hole limit of ST theories assuming that the sensitivity for the stationary black-hole holds for a binary system. The sensitivity parameter (s_A) for the stationary black-hole in the Jordan-Frame conventions is exactly $s_A = \frac{1}{2}$ (see, Refs. [30, 31]), which in our conventions[5] implies that the parameter $\alpha_A = 0$, and hence all other ST parameters entering the scattering angle vanish (see, Table 2.1). This shows that our results are indistinguishable from the results of GR under this assumption.

This work must be seen as a first step to compute the gauge-invariant scattering angle within the PN expansion for massless scalar-tensor theories. These results can also be used in future as a test for the PN expansion of the post-Minkowskian (PM) scattering angle computed for ST theories using the PM effective-field-theory formalism.

References

1. Iwasaki Y (1971) Quantum theory of gravitation versus classical theory. - fourth-order potential. Prog Theor Phys 46:1587–1609. https://doi.org/10.1143/PTP.46.1587
2. Goldberger WD, Rothstein IZ (2006) An effective field theory of gravity for extended objects. Phys Rev D 73:104029. arXiv:hep-th/0409156. https://doi.org/10.1103/PhysRevD.73.104029
3. Kälin G, Porto RA (2020) Post-Minkowskian effective field theory for conservative binary dynamics. JHEP 11:106. arXiv:2006.01184. https://doi.org/10.1007/JHEP11(2020)106 [hep-th]
4. Kälin G, Porto RA (2019) From boundary data to bound states. JHEP 01:072. arXiv:1910.03008. https://doi.org/10.1007/JHEP01(2020)072 [hep-th]
5. Damour T (2016) Gravitational scattering, post-Minkowskian approximation and effective one-body theory. Phys Rev D94:104015. arXiv:1609.00354. https://doi.org/10.1103/PhysRevD.94.104015 [gr-qc]

[5] The Jordan-Frame sensitivity (s_A) is related to the parameter α_A as,

$$s_A = \frac{1}{2} - \frac{\alpha_A}{2\alpha_0}. \tag{4.73}$$

6. Mukherjee S, Mitra S, Chatterjee S (2021) Gravitational wave observatories may be able to detect hyperbolic encounters of black holes. Mon Not Roy Astron Soc 508:5064–5073. arXiv:2010.00916. https://doi.org/10.1093/mnras/stab2721 [gr-qc]
7. Gold R, Brügmann B (2013) Eccentric black hole mergers and zoom-whirl behavior from elliptic inspirals to hyperbolic encounters. Phys Rev D 88:064051. arXiv:1209.4085. https://doi.org/10.1103/PhysRevD.88.064051 [gr-qc]
8. Jain T (2023) Gravitational scattering up to third post-Newtonian approximation for conservative dynamics: scalar-tensor theories. Phys Rev D 108:104071. arXiv:2304.09052. https://doi.org/10.1103/PhysRevD.108.104071 [gr-qc]
9. Bini D, Damour T (2017) Gravitational scattering of two black holes at the fourth post-Newtonian approximation. Phys Rev D96:064021. arXiv:1706.06877. https://doi.org/10.1103/PhysRevD.96.064021 [gr-qc]
10. Damour T, Esposito-Farese G (1992) Tensor multiscalar theories of gravitation. Class Quant Grav 9:2093–2176. https://doi.org/10.1088/0264-9381/9/9/015
11. Damour T, Esposito-Farese G (1996) Testing gravity to second postNewtonian order: a field theory approach. Phys Rev D 53:5541–5578. arXiv:gr-qc/9506063. https://doi.org/10.1103/PhysRevD.53.5541
12. Damour T, Jaranowski P, Schaefer G (2000) On the determination of the last stable orbit for circular general relativistic binaries at the third postNewtonian approximation. Phys Rev D62:084011. arXiv:gr-qc/0005034. https://doi.org/10.1103/PhysRevD.62.084011 [gr-qc]
13. Bernard L (2018) Dynamics of compact binary systems in scalar-tensor theories: equations of motion to the third post-Newtonian order. Phys Rev D 98:044004. arXiv:1802.10201. https://doi.org/10.1103/PhysRevD.98.044004 [gr-qc]
14. Bernard L (2019) Dynamics of compact binary systems in scalar-tensor theories: II. Center-of-mass and conserved quantities to 3PN order. Phys Rev D 99:044047. arXiv:1812.04169. https://doi.org/10.1103/PhysRevD.99.044047 [gr-qc]
15. Damour T, Jaranowski P, Schäfer G (2015) Fourth post-Newtonian effective one-body dynamics. Phys Rev D 91:084024. arXiv:1502.07245. https://doi.org/10.1103/PhysRevD.91.084024 [gr-qc]
16. Jain T, Rettegno P, Agathos M, Nagar A, Turco L (2023) Effective-one-body Hamiltonian in scalar-tensor gravity at third post-Newtonian order. Phys. Rev. D 107:084017. arXiv:2211.15580. https://doi.org/10.1103/PhysRevD.107.084017 [gr-qc]
17. Julié F-L, Deruelle N (2017) Two-body problem in Scalar-Tensor theories as a deformation of general relativity : an effective-one-body approach. Phys Rev D 95:124054. arXiv:1703.05360. https://doi.org/10.1103/PhysRevD.95.124054 [gr-qc]
18. Julié F-L (2018) Reducing the two-body problem in scalar-tensor theories to the motion of a test particle : a scalar-tensor effective-one-body approach. Phys Rev D 97:024047. arXiv:1709.09742. https://doi.org/10.1103/PhysRevD.97.024047 [gr-qc]
19. Jain T (2023) Nonlocal-in-time effective one body Hamiltonian in scalar-tensor gravity at third post-Newtonian order. Phys Rev D 107:084018. arXiv:2301.01070. https://doi.org/10.1103/PhysRevD.107.084018 [gr-qc]
20. Julié F-L, Baibhav V, Berti E, Buonanno A (2023) Third post-Newtonian effective-one-body Hamiltonian in scalar-tensor and Einstein-scalar-Gauss-Bonnet gravity. Phys Rev D 107:104044. arXiv:2212.13802. https://doi.org/10.1103/PhysRevD.107.104044 [gr-qc]
21. Lang RN (2014) Compact binary systems in scalar-tensor gravity. II. Tensor gravitational waves to second post-Newtonian order. Phys Rev D 89:084014. arXiv:1310.3320. https://doi.org/10.1103/PhysRevD.89.084014 [gr-qc]
22. Lang RN (2015) Compact binary systems in scalar-tensor gravity. III. Scalar waves and energy flux. Phys Rev D 91:084027. arXiv:1411.3073. https://doi.org/10.1103/PhysRevD.91.084027 [gr-qc]
23. Landau L, Lifshitz E (1976) Mechanics, 3rd edn, vol 1. Course of theoretical physics. Elsevier Science
24. Damour T, Schäfer G (1988) Higher order relativistic periastron advances and binary pulsars. Nuovo Cim B101:127. https://doi.org/10.1007/BF02828697

25. Blanchet L, Faye G (2000) Hadamard regularization. J Math Phys 41:7675–7714. arXiv:gr-qc/0004008. https://doi.org/10.1063/1.1308506
26. Gradshteyn IS, Ryzhik IM (2007) Table of integrals, series, and products, 7th edn Elsevier/Academic Press, Amsterdam, pp xlviii+1171, translated from the Russian, Translation edited and with a preface by Alan Jeffrey and Daniel Zwillinger, With one CD-ROM (Windows, Macintosh and UNIX)
27. Bini D, Geralico A (2021) Frequency domain analysis of the gravitational wave energy loss in hyperbolic encounters. Phys Rev D 104:104019. arXiv:2108.02472. https://doi.org/10.1103/PhysRevD.104.104019 [gr-qc]
28. Damour T, Jaranowski P, Schäfer G (2014) Nonlocal-in-time action for the fourth post-Newtonian conservative dynamics of two-body systems. Phys Rev D 89:064058. arXiv:1401.4548. https://doi.org/10.1103/PhysRevD.89.064058 [gr-qc]
29. Abramowitz M (1974) Handbook of mathematical functions, with formulas, graphs, and mathematical tables. Dover Publications, Inc., USA
30. Hawking S, Hartle J (1972) Energy and angular momentum flow into a black hole. Commun Math Phys 27:283–290. https://doi.org/10.1007/BF01645515
31. Sotiriou TP, Faraoni V (2012) Black holes in scalar-tensor gravity. Phys Rev Lett 108:081103. arXiv:1109.6324. https://doi.org/10.1103/PhysRevLett.108.081103 [gr-qc]

Chapter 5
Angular Momentum Flux in Scalar-Tensor Theories up to 1.5 Post-Newtonian Order

We compute the angular momentum flux from an inspiralling nonspinning binary system of compact objects in massless scalar-tensor (ST) theories. The angular momentum flux in scalar-tensor theories involves both a tensorial and a scalar contribution which can be further decomposed as instantaneous, tail and a non-linear memory effect up to 1.5PN order. Here, we compute the angular momentum flux up to 1.5PN order for the tensorial part and 1PN order for the scalar part using the tensorial and the scalar multipole moments. This work is important to construct the radiation-reaction force and hence the waveform templates for eccentric binaries in scalar-tensor theories of gravity.

Before going into the details of our calculation, let us first briefly discuss the hereditary contributions, i.e., the contributions that depend on the entire past history of the system. These hereditary contributions entering in the waveform has two-fold effect in the form of tail effects and memory effects. The physical origin of the tail effects can be understood from the scattering of linear gravitational waves off the background curvature generated by the total mass of the system. A second contribution, which enters at the same PN order, known as the non-linear memory effect could be understood as the re-radiation of GWs by the stress-energy distribution (tensor) of GW themselves [1–4].

However, it should be noted that in GR, there is also a second type of memory effect called the linear memory effect. The linear memory effect arises from the sources that produce a net change in the time derivatives of one or more of its source-multipole moments, for example, hyperbolic encounters where the memory arises due to the difference in the direction of the velocity vector at early and late times. The bound orbit sources like the ones currently detected by GW detectors generally have a vanishing linear memory effect but a non-vanishing non-linear memory effect. The non-linear memory effect arises from a contribution to the radiative multipole moment which is sourced by the GW stress energy tensor. These radiative multipolar moments are related to the source multipole moments using the multipolar-post-Minkowskian (MPM) approach [4].

This is based on the work in Ref. [5].

T. Jain, *Surfing the Gravitational Waves*, Springer Theses,
https://doi.org/10.1007/978-3-031-96180-9_5

5.1 Introduction

The equations-of-motion (EOM) and the multipole moments for the nonspinning two-body problem in scalar-tensor theories are known up to 3PN order and 1.5PN order, respectively [6–12]. This implies that the energy and angular momentum flux can be computed up to 1.5PN order for inspiralling compact binaries moving in general *non-circular* orbits. The contributions up to 1.5PN order to both energy and angular momentum fluxes in scalar-tensor theories come from the instantaneous terms and the (non-linear) hereditary terms [12]. The energy flux for the non-circular orbits in scalar-tensor theories is computed in Ref. [12].

In this work, we compute the angular momentum flux of inspiralling nonspinning compact binaries up to 1.5PN order. This could then be used to compute the radiation-reaction force and hence the waveform templates for eccentric binaries in scalar-tensor theories of gravity.

This Chapter is organised as follows. In Sect. 5.2, we give a brief overview of scalar-tensor (ST) parameters and define the shorthand notations used for ST parameters in this chapter followed by the general structure of angular momentum flux in ST theories in Sect. 5.3. We compute the tensorial contribution to the angular momentum flux including the instantaneous, tail and memory effects up to 1.5PN order in Sect. 5.4 and then the instantaneous and tail contributions due to the scalar in Sect. 5.5. We will also consider geometric units, such as $G = c = 1$ unless specified explicitly.

5.2 Scalar-Tensor Parameters and Conventions

In this Chapter as well, we consider single scalar massless ST theories in which a scalar field φ is minimally coupled to the metric $g_{\mu\nu}$ in the Einstein frame. This theory is described (in the Einstein frame) by the action

$$S = \frac{c^4}{16\pi G}\int d^4x\sqrt{-g}(R - 2g^{\mu\nu}\partial_\mu\varphi\partial_\nu\varphi) + S_m[\Psi, \mathscr{A}(\varphi)^2 g_{\mu\nu}], \tag{5.1}$$

where $g_{\mu\nu}$ is the Einstein metric, R is the Ricci scalar, φ is the scalar field, Ψ collectively denotes the matter fields, $g \equiv \det(g_{\mu\nu})$ and G is the Newtonian constant.

Here too we use the conventions of Refs. [13–15], i.e. DEF conventions, to define the body-dependent ST parameters that encompass the effects due to the scalar field in the system's dynamics. The mass function $m_J(\varphi)$ in the Einstein frame is used to define these dimensionless body-dependent parameters in DEF conventions (see, Eqs. (2.6)–(2.9) in Chap. 2 for the definition of the parameters).

The scalar and the tensorial source multipole moments derived in Ref. [12] are expressed in terms of the Jordan frame ST parameters. The conversion of the Jordan-Frame parameters to DEF conventions, i.e. the Einstein-Frame parameters, is given in Table 2.1 of Chap. 2. Along with the definitions in Table 2.1, we also use the following combinations of the asymptotic scalar parameters,

$$\mathcal{S}_A = \sqrt{\frac{1+\alpha_0^2}{1+\alpha_A^0\alpha_B^0}}\left(\frac{\alpha_A^0}{\alpha_0}\right), \tag{5.2a}$$

$$\theta_A = -\frac{\alpha_B^0}{4\alpha_A^0}\frac{\beta_A^0\mathcal{S}_A}{\left(1+\alpha_A^0\alpha_B^0\right)}, \tag{5.2b}$$

$$\Theta_A = \frac{\alpha_B\beta_A')^0\mathcal{S}_B}{8\left(1+\alpha_A^0\alpha_B^0\right)^2}, \tag{5.2c}$$

together with the corresponding ones obtained by substituting $A \leftrightarrow B$. Here, α_A^0, α_0 and β_A^0 denote the asymptotic value of the ST parameters defined in Eqs. (2.6), (2.7) and (2.5), e.g. $\alpha_A^0 \equiv \alpha_A(\varphi_0)$ with φ_0 being the asymptotic value of the scalar field far from the two bodies. The (anti-)symmetric combinations of these quantities are defined as usual as,

$$x_\pm = \frac{(x_A \pm x_B)}{2}. \tag{5.3}$$

Finally, we will also make use of the following combinations defined in Chap. 2,

$$\langle\bar{\beta}\rangle = \bar{\beta}_+ - X_{AB}\bar{\beta}_-, \tag{5.4a}$$

$$\langle\delta\rangle = \delta_+ + X_{AB}\delta_-, \tag{5.4b}$$

along with some additional combinations of the ST coefficients,

$$\zeta' = 1 - \alpha_{AB}\frac{2+\gamma_{AB})}{2} = \frac{\alpha_0^2}{1+\alpha_0^2} \tag{5.4c}$$

$$\langle\theta\rangle = \theta_+ - X_{AB}\theta_-, \tag{5.4d}$$

$$\langle\Theta\rangle = \Theta_+ - X_{AB}\Theta_-. \tag{5.4e}$$

5.3 The Far-Zone Angular Momentum Flux in Scalar-Tensor Theories

In this section, we discuss the basic structure of the angular momentum flux of inspiralling, nonspinning compact binaries in ST theory, following the work of Ref. [12], which computed the energy flux up to 1.5PN order. We here define the n-th PN order (following Ref. [12]) as the term proportional to $1/c^{2n}$ beyond the leading GR quadrupole term. This means that the leading-order ST dipolar term will be denoted as -1PN order.

The angular momentum flux in ST theories can be expressed as the sum of a tensorial and a scalar contribution, such that

$$\mathcal{G}_i = \mathcal{G}_i^{\mathrm{T}} + \mathcal{G}_i^{\mathrm{S}} \equiv \left(\frac{\mathrm{d}L_i}{\mathrm{d}t}\right)^{\mathrm{GW}}, \tag{5.5}$$

where the superscripts "S" and "T" denote the scalar and tensorial contributions respectively.

The explicit expressions of angular momentum flux are derived using the radiative multipole moments. We will compute the tensorial angular momentum flux $\mathcal{G}_i^{\mathrm{T}}$ up to 1.5PN order, including both instantaneous and hereditary contributions (entering at 1.5PN order). We compute the scalar flux $\mathcal{G}_i^{\mathrm{S}}$ explicitly only up to 1PN, since for the instantaneous contribution at 1.5PN order one needs the explicit expression of the scalar dipole moments at relative 2.5PN accuracy which are not explicitly given in Ref. [12].

We describe two particles with relative separation x^{i} and relative velocity v^{i}. We will use dots to denote time derivatives, e.g. $\mathrm{v}^i = \dot{\mathrm{x}}^i \equiv \mathrm{dx}^i/\mathrm{d}t$. The angular momentum of the system is then computed as $L_i = \varepsilon_{ijk}\mathrm{x}^{\mathrm{j}}\mathrm{v}^{\mathrm{k}}$. When working with polar coordinates (r, ϕ), we define the unit direction along the angular momentum as

$$\hat{L}_i \equiv \frac{\varepsilon_{ijk}\mathrm{x}^{\mathrm{j}}\mathrm{v}^{\mathrm{k}}}{r^2\dot{\phi}}, \tag{5.6}$$

and the total velocity of the system as $\mathrm{v} = \sqrt{\dot{r}^2 + r^2\dot{\phi}^2}$.

The angular momentum flux $\mathcal{G}_i$ is orthogonal to the orbital plane, and aligned with the angular momentum L_i. We can then compute the modulus of the angular momentum flux $\mathcal{G}$, such that

$$\mathcal{G}_i = \mathcal{G}\hat{L}_i. \tag{5.7}$$

5.4 Tensorial Contribution

The tensorial contribution to the angular momentum flux up to 1.5PN is expressed through the formulas derived in Ref. [16], namely

$$\mathcal{G}_i^{\mathrm{T}} = \frac{G\phi_0^{\mathrm{SF}}}{c^5}\varepsilon_{ijk}\left\{\frac{2}{5}\mathcal{U}_{j\ell}\,\dot{\mathcal{U}}_{k\ell} + \frac{1}{c^2}\left[\frac{1}{63}\mathcal{U}_{j\ell m}\,\dot{\mathcal{U}}_{k\ell m} + \frac{32}{45}\mathcal{V}_{j\ell}\,\dot{\mathcal{V}}_{k\ell}\right] + O\left(\frac{1}{c^4}\right)\right\}, \tag{5.8}$$

where $\mathcal{U}$ and $\mathcal{V}$ are, respectively, the symmetric-trace-free mass and current radiative tensorial multipole moments. Only the mass quadrupole multipole moment, $\mathcal{U}_{j\ell}$, differs from its twice differentiated source counterpart (I_{jk}) up to 1.5PN order (see, Refs. [8, 9, 12]) by

$$\mathscr{U}_{j\ell} = \ddot{I}_{j\ell} + \frac{2GM}{\phi_0^{\rm SF} c^3} \int_{-\infty}^{T_{\rm R}} dV \overset{(4)}{I}_{j\ell}(V) \left[\ln\left(\frac{T_{\rm R} - V}{2\tau_0} \right) + \frac{11}{12} \right] + \frac{G(1-\zeta')}{3\zeta' c^3} \left(\int_{-\infty}^{T_{\rm R}} dV \ddot{I}^{\rm S}_{\langle j}(V) \ddot{I}^{\rm S}_{\ell\rangle}(V) - A_{j\ell} \right) + O\left(\frac{1}{c^4}\right), \tag{5.9}$$

where

$$A_{j\ell} = \overset{(3)}{I}{}^{\rm S}_{\langle j} I^{\rm S}_{\ell\rangle} + \dot{I}^{\rm S}_{\langle j} \ddot{I}^{\rm S}_{\ell\rangle} + \frac{1}{2} I^{\rm S} \overset{(3)}{I}{}^{\rm S}_{j\ell}, \tag{5.10}$$

and we have introduced the superscripts "(n)" to denote n-th time derivatives. Hereafter, the angular brackets denote the symmetric trace-free (STF) part. The quadrupole radiative moment $\mathscr{U}_{ij}$ given in Eq. (5.9) is a combination of two contributions: (i) an instantaneous contribution, which is a function of retarded time $T_{\rm R}$, and (ii) a hereditary contribution, which depends on the dynamics of the system in the entire past, i.e. $V \leq T_{\rm R}$. The parameter τ_0 is a constant time scale that enters the relation between the retarded time $T_{\rm R}$ in radiative coordinates and the corresponding time in harmonic coordinates. Therefore, from the above equation, we observe that, starting at 1.5PN order, the tensorial part of the angular momentum flux also contains both instantaneous and hereditary (tail and memory) contributions, so that we can write

$$\mathscr{G}^{\rm T} = \mathscr{G}^{\rm inst,T} + \mathscr{G}^{\rm hered,T}. \tag{5.11}$$

The hereditary contributions can then also be further decomposed as,

$$\mathscr{G}_i^{\rm T,hered} = \mathscr{G}_i^{\rm T,tail} + \mathscr{G}_i^{\rm T,memory}, \tag{5.12}$$

where $\mathscr{G}_i^{\rm T,memory}$ is the non-linear memory effect and $\mathscr{G}_i^{\rm T,tail}$ is the tail effect. We will derive the explicit expressions for these contributions in the following subsections.

5.4.1 Instantaneous Contribution

The instantaneous tensorial term entering the angular momentum flux up to 1.5PN order in terms of the source multipole moments following Eq. (5.8) is,

$$\mathscr{G}^{\rm inst,T} = \frac{G\phi_0^{\rm SF}}{c^5} \varepsilon_{ijk} \left\{ \frac{2}{5} \ddot{I}_{ja} \overset{(3)}{I}_{ka} + \frac{1}{c^2} \left(\frac{1}{63} \overset{(3)}{I}_{jab} \overset{(4)}{I}_{kab} + \frac{32}{45} \ddot{J}_{ja} \overset{(3)}{J}_{ka} \right) - \frac{2G(1-\zeta')}{15\zeta' c^3} \left(A_{ja} \overset{(3)}{I}_{ka} + \dot{A}_{ka} \ddot{I}_{ja} \right) + \frac{2G(1-\zeta')}{15\zeta' c^3} \ddot{I}_{ja} \ddot{I}^{\rm S}_{\langle k} \ddot{I}^{\rm S}_{a\rangle} \right\}. \tag{5.13}$$

Here, I_{ij} and I_{ijk} are the tensorial source quadrupole and octupole multipole moments and I^S, I^S_i and I^S_{ij} are the scalar source monopole, dipole and quadrupole moments, respectively (see, Ref. [12] for their explicit expressions).

This instantaneous tensorial term entering the angular momentum flux up to 1.5PN order can be written as,

$$\mathscr{G}^{\text{inst,T}} = \mathscr{G}^{\text{inst,T}}_{\text{N}} + \mathscr{G}^{\text{inst,T}}_{0.5\text{PN}} + \mathscr{G}^{\text{inst,T}}_{1\text{PN}} + \mathscr{G}^{\text{inst,T}}_{1.5\text{PN}}. \tag{5.14}$$

Using multipole moments in Eq. (5.13), the contributions up to 1.5PN order in harmonic coordinates in DEF conventions are found to be

$$\mathscr{G}^{\text{inst,T}}_{\text{N}} = \frac{G_{AB}^2 M^3 \nu^2}{c^5 r}\dot{\phi}\left(1 + \frac{\gamma_{AB}}{2}\right)\frac{8}{5}\left(2\frac{G_{AB}M}{r} + 2\text{v}^2 - 3\dot{r}^2\right), \tag{5.15a}$$

$$\mathscr{G}^{\text{inst,T}}_{0.5\text{PN}} = 0, \tag{5.15b}$$

$$\begin{aligned}
\mathscr{G}^{\text{inst,T}}_{1\text{PN}} &= \frac{G_{AB}^2 M^3 \nu^2}{c^7 r}\dot{\phi}\left(1 + \frac{\gamma_{AB}}{2}\right)\frac{8}{5}\left\{\left(-\frac{745}{42} + \frac{\nu}{21} - 7\gamma_{AB} - 8\langle\bar{\beta}\rangle\right)\left(\frac{G_{AB}M}{r}\right)^2\right. \\
&+ \left(-\frac{58}{21} - \frac{95}{21}\nu - 4\langle\bar{\beta}\rangle\right)\frac{G_{AB}M}{r}\text{v}^2 + \left(\frac{62}{7} + \frac{197}{42}\nu + 2\gamma_{AB} + 8\langle\bar{\beta}\rangle\right)\frac{G_{AB}M}{r}\dot{r}^2 \\
&\left. + \left(\frac{307}{84} - \frac{137}{21}\nu + \gamma\right)\text{v}^4 + \left(-\frac{37}{7} + \frac{277}{14}\nu\right)\text{v}^2\dot{r}^2 + \left(\frac{95}{28} - \frac{90}{7}\nu\right)\dot{r}^4\right\},
\end{aligned} \tag{5.15c}$$

$$\begin{aligned}
\mathscr{G}^{\text{inst,T}}_{1.5\text{PN}} &= \frac{G_{AB}^2 M^3 \nu^2}{c^8 r}\dot{\phi}\left(1 + \frac{\gamma_{AB}}{2}\right)\left(\frac{8}{5}\zeta'\frac{G_{AB}M}{r}\dot{r}\right)\left\{\left[\mathscr{S}_+^2 - \left(1 - \frac{50}{3}\nu\right)\mathscr{S}_-^2\right]\frac{G_{AB}M}{r}\right. \\
&\left. + 2\left[\mathscr{S}_+^2 - (1 - 14\nu)\mathscr{S}_-^2\right]\text{v}^2 - \frac{5}{2}\left[\mathscr{S}_+^2 + (1 - 16\nu)\mathscr{S}_-^2\right]\dot{r}^2\right\}.
\end{aligned} \tag{5.15d}$$

The above results also contain an instantaneous contribution coming from the memory term of Eq. (5.9) entering the angular momentum flux. The memory term of Eq. (5.9) is just an anti-derivative term, therefore, one of its contributions corresponding to the time derivative of $\mathscr{U}_{kl}$ in Eq. (5.8) becomes instantaneous. This contribution enters at 1.5PN order and hence, is incorporated in the 1.5PN instantaneous contribution given above (see, Eq. (5.15d)). There is also a remaining memory effect entering at 1.5PN order which we will discuss in Sect. 5.4.2.1. In contrast, there are no memory contributions in the energy flux as in the energy flux formula the memory is completely time-differentiated and therefore becomes instantaneous (see, Ref. [12]).

The results obtained in this section match the angular momentum flux computed in Refs. [17–20] in the GR limit, i.e. $\alpha^0_{A,B} \to 0$.

5.4.2 Hereditary Contributions

Now, we move on to compute the hereditary contribution to the tensorial part of the angular momentum flux. As for the flux computations in GR [19, 20], it is not possible to obtain closed-form results for these contributions in the time-domain, therefore we adopt the same strategy used in GR of using the discrete Fourier decomposition in scalar-tensor theory.

As mentioned previously, the hereditary contribution can be further decomposed as,

$$\mathscr{G}_i^{\mathrm{hered,T}} = \mathscr{G}_i^{\mathrm{tail,T}} + \mathscr{G}_i^{\mathrm{mem,T}} \tag{5.16}$$

Combining Eqs. (5.8) and (5.9), the tail and memory integrals entering in the tensorial angular momentum flux up to 1.5PN order in ST theory are given by

$$\begin{aligned}\mathscr{G}_i^{\mathrm{T,tail}} = \frac{4G^2M}{5c^8}\varepsilon_{ijk}\Bigg\{&\ddot{I}_{j\ell}\int_{-\infty}^{T_\mathrm{R}} dV \overset{(5)}{I}_{k\ell}(V)\left[\ln\left(\frac{T_\mathrm{R}-V}{2\tau_0}\right)+\frac{11}{12}\right]\\ &-\overset{(3)}{I}_{j\ell}\int_{-\infty}^{T_\mathrm{R}} dV \overset{(4)}{I}_{k\ell}(V)\left[\ln\left(\frac{T_\mathrm{R}-V}{2\tau_0}\right)+\frac{11}{12}\right]\Bigg\}\end{aligned} \tag{5.17}$$

and

$$\mathscr{G}_i^{\mathrm{T,mem}} = \frac{2G^2(1-\zeta')\phi_0^{\mathrm{SF}}}{15\zeta' c^8}\varepsilon_{ijk} I_{j\ell}^{(3)} \int_{-\infty}^{T_\mathrm{R}} dV \ddot{I}^{\mathrm{S}}_{\langle k}(V)\ddot{I}^{\mathrm{S}}_{\ell\rangle}(V). \tag{5.18}$$

From Eq. (5.17), we observe that the leading order tail contribution entering at 1.5PN order is the usual GR tail contribution [19–22] and the corrections due to the scalar-field only enter as a memory effect at 1.5PN order (see, Eq. (5.18) for its explicit expression). Therefore, in this section, we will focus only on computing the memory effect contribution.

5.4.2.1 Memory Contribution

The memory contribution to the tensorial part of the angular momentum flux from Eq. (5.18) is given by,

$$\mathscr{G}_i^{\mathrm{T,mem}} = \frac{2G^2(1-\zeta')\phi_0^{\mathrm{SF}}}{15\zeta' c^8}\varepsilon_{ijk} \overset{(3)}{I}_{j\ell}(\mathrm{T_R}) \int_{-\infty}^{T_\mathrm{R}} dV \ddot{I}^{\mathrm{S}}_{k}(V)\ddot{I}^{\mathrm{S}}_{\ell}(V). \tag{5.19}$$

The above equation is symmetric with respect to the indices k and ℓ and is also trace-free due to the presence of ε_{ijk}, hence we omit the angular brackets from the indices k and ℓ that indicate the STF projection. The orbital average therefore reads,

$$\langle\mathscr{G}_i^{\mathrm{T,mem}}\rangle = \frac{2G^2(1-\zeta')\phi_0^{\mathrm{SF}}}{15\zeta' c^8}\varepsilon_{ijk}\int_0^P \frac{dT_\mathrm{R}}{P}\overset{(3)}{I}_{j\ell}(\mathrm{T_R}) \int_{-\infty}^{T_\mathrm{R}} dV \ddot{I}^{\mathrm{S}}_{k}(V)\ddot{I}^{\mathrm{S}}_{\ell}(V), \tag{5.20}$$

which can be re-expressed as,

$$\langle\mathscr{G}_i^{\mathrm{T,mem}}\rangle = \frac{2G^2(1-\zeta')\phi_0^{\mathrm{SF}}}{15\zeta' c^8}\varepsilon_{ijk}\Bigg\{\int_{-\infty}^{0} dV \ddot{I}^{\mathrm{S}}_{k}(V)\ddot{I}^{\mathrm{S}}_{\ell}(V)\int_0^P \frac{dT_\mathrm{R}}{P}\overset{(3)}{I}_{j\ell}(\mathrm{T_R})$$

$$+\int_0^P dV \ddot{I}_k^S(V)\ddot{I}_\ell^S(V)\int_V^P \frac{dT_R}{P}\overset{(3)}{I}_{j\ell}(T_R)\Bigg\} \tag{5.21}$$

At first stance, the contribution due to the first term appears to be zero as this term contains the orbital average of the third time-derivative of the Newtonian tensorial quadrupole moment up to 1.5PN order and since the motion is periodic, the quadrupole averages to zero. However, as mentioned in Ref. [20], we cannot replace this term with zero because this argument neglects the evolution of the Keplerian orbital elements in the remote past. We study the impact of this term and compute the non-zero DC contribution to the memory effects in scalar-tensor theory.

In contrast, the contribution of the second term, i.e. due to the recent past of the source averages to zero. This implies that the non-zero DC contribution is due to the remote past of the source, i.e., the first term of Eq. (5.21).

Now we compute the DC contribution of the non-linear memory integral following the method of Ref. [20]. For the leading order contribution to the memory integral, i.e. at 1.5PN order, it is sufficient to replace the tensorial quadrupolar moment and scalar dipole moment and their time derivatives with their Newtonian values. Here, we work in polar coordinates (r, ϕ) and express Eq. (5.19) in terms of the current orbital separation r_0, radial velocity $\dot{r}_0$, orbital frequency $\dot{\phi}_0$ and orbital phase $\phi_0 = 0$. For any instant in the past, we will use the notation r, $\dot{r}$ and $\dot{\phi}$ to denote the orbital separation, radial velocity and orbital frequency, respectively. From this, we obtain

$$\mathcal{G}^{\text{mem}} = -\frac{8\nu^3}{15r_0}\frac{G_{AB}^5 M^6 \mathcal{S}_-^2(1-\zeta')\zeta'}{\alpha_{AB}}\int_{t_1}^{T_R} dV\left(4\frac{\dot{\phi}_0}{r^4}\cos(2\phi) - \frac{\dot{r}_0}{r_0 r^4}\sin(2\phi)\right), \tag{5.22}$$

where we denote by t_1 some initial instant of the formation of the system with some eccentricity close to 1.

We then replace r, $\dot{r}$ and $\dot{\phi}$ in the above equation by the explicit solution for Keplerian motion in ST theories, i.e.,

$$r = \frac{a(1-e^2)}{1+e\cos\phi}, \tag{5.23}$$

$$\dot{r} = \sqrt{\frac{G_{AB}}{a(1-e^2)}}e\sin\phi, \tag{5.24}$$

$$\dot{\phi} = \sqrt{\frac{G_{AB}}{a^3(1-e^2)^3}}(1+e\cos\phi)^2. \tag{5.25}$$

On inserting the above expressions in Eq. (5.22), we obtain an integrand as a function of the semi-major axis $a(t)$, eccentricity $e(t)$ and orbital phase $\phi(t)$ such that the resulting integral is composed of many harmonics of the Keplerian motion in terms of the orbital phase $\phi(t)$. The orbital phase oscillates at the rate of the orbital period in comparison to $a(t)$ and $e(t)$ which remain approximately constant during one period but slowly evolve via radiation reaction during the past evolution of the binary. We expect that the oscillations of the orbital phase, due to the sequence of many orbital cycles in the entire life of the the system will more or less cancel each other in the memory integral so that the true PN order of the oscillating terms will be given by the power of $1/c$ they carry.

The explicit expression then becomes,

$$\mathscr{G}^{\rm mem} = -\frac{8\nu^3\dot{\phi}_0}{15r_0}\frac{G_{AB}^5M^6\mathscr{S}_-^2(1-\zeta')\zeta'}{\alpha_{AB}}\int_{t_1}^{t_0}dt\frac{e^2(e^2+6)}{a^4(e^2-1)^4}. \tag{5.26}$$

5.5 Scalar Contribution

To derive the scalar contribution to the angular momentum flux in massless ST theories, we first need to find its relation to the radiative scalar multipole moments. The corresponding formula, linking the scalar contribution to the *energy* flux in terms of radiative multipole moments is given in Eq. (3.27b) of Ref. [12].

The angular momentum losses can be easily calculated using the energy-momentum tensor for scalar radiation. The scalar waveform ψ for massless ST theory is given by Eq. (3.25) of Ref. [12] and reads

$$\psi = -\frac{2G}{c^2R}\sum_{\ell=0}^{\infty}\frac{1}{c^\ell \ell!}N_{i_1i_2\cdots i_\ell}\,\mathscr{U}^{\rm S}_{i_1i_2\cdots i_\ell}, \tag{5.27}$$

where $\mathscr{U}^{\rm S}_{i_1i_2\cdots i_\ell}$ is the scalar radiative moment, and $N_{i_1i_2\cdots i_\ell} \equiv N_{i_1}\cdots N_{i_\ell}$. N_i is a vector of the polarization orthonormal triad $(\mathbf{N}, \mathbf{P}, \mathbf{Q})$ [12, 23]. The energy-momentum tensor for the scalar GWs corresponding to the action of Ref. [12] is,

$$T^{\rm S,GW}_{\alpha\beta} = \frac{c^3\phi_0^{\rm SF}(1-\zeta')}{16\zeta'\pi G}\langle\partial_\alpha\psi\,\partial_\beta\psi\rangle. \tag{5.28}$$

Hence, the scalar angular momentum flux in ST theory is (see, Ref. [16] for GR)

$$\begin{aligned}\mathscr{F}_J^{\rm S} &= -\int R^2\varepsilon_{ijk}x_jT^{\rm S,GW}_{0k}\,d\Omega\\ &= -\frac{c^3R^2\phi_0^{\rm SF}(1-\zeta')}{16\zeta'\pi G}\int\varepsilon_{ijk}x_j\partial_0\psi\,\partial_k\psi\,d\Omega.\end{aligned} \tag{5.29}$$

Now using,

$$\partial_0\psi = -\frac{2G}{Rc^2}\sum_{\ell=0}^{\infty}\frac{1}{c^\ell \ell!}N_{i_1i_2\cdots i_\ell}\,\dot{\mathscr{U}}^{\rm S}_{i_1i_2\cdots i_\ell}, \tag{5.30}$$

$$\partial_k \psi = -\frac{2G}{R^2c^2}\sum_{\ell=0}^{\infty}\frac{\ell+1}{c^{\ell+1}(\ell+1)!}N_{i_1i_2\cdots i_\ell}\,\mathscr{U}^{\rm S}_{ki_1i_2\cdots i_\ell}$$
$$+\frac{2G}{R^2c^2}\sum_{\ell=0}^{\infty}\frac{n_k}{c^{\ell+1}(\ell+1)!}N_{i_1i_2\cdots i_{\ell+1}}\,\mathscr{U}^{\rm S}_{i_1i_2\cdots i_{\ell+1}}, \tag{5.31}$$

in Eq. (5.29), we obtain

$$\mathscr{F}_J^{\rm S} = -\frac{G\phi_0^{\rm SF}(1-\zeta')}{4\zeta'\pi c}\varepsilon_{ijk}n_j\int d\Omega\sum_{\ell=0}^{\infty}\sum_{\ell'=0}^{\infty}\frac{1}{c^{\ell}c^{\ell'+1}\ell!(\ell'+1)!}\Big[(\ell'+1)N_{A_l}N_{B_{\ell'}}\dot{\mathscr{U}}^{\rm S}_{A_\ell}\dot{\mathscr{U}}^{\rm S}_{kB_{\ell'}}$$
$$-n_kN_{A_\ell}N_{B_{\ell'+1}}\dot{\mathscr{U}}^{\rm S}_{A_\ell}\dot{\mathscr{U}}^{\rm S}_{B_{\ell'+1}}\Big]. \tag{5.32}$$

Integrating the above equation using Eq. (2.6) of Ref. [16], we find that the only non-zero contribution comes from the first term in the bracket when $l' = l - 1$, i.e.

$$\mathscr{F}_J^{\rm S} = \varepsilon_{ijk}\sum_{l=1}^{\infty}\frac{G\phi_0^{\rm SF}(1-\zeta')}{\zeta' c^{2l+1}(l-1)!(2l+1)!!}\dot{\mathscr{U}}^{\rm S}_{ji_1i_2\cdots i_{l-1}}\,\mathscr{U}^{\rm S}_{ki_1i_2\cdots i_{l-1}} \tag{5.33}$$

Substituting the expression of the scalar radiative moment from Ref. [12] in the above equation, we obtain the scalar contribution to the angular momentum flux. Note that, similar to the tensorial contribution to the angular momentum flux, the scalar contribution also contains both instantaneous and tail contributions, so that we can write

$$\mathscr{G}^{\rm S} = \mathscr{G}^{\rm inst,S} + \mathscr{G}^{\rm tail,S}. \tag{5.34}$$

This is because the monopole scalar radiative moment $\mathscr{U}_j^{\rm S}$ entering Eq. (5.33) differs from its time differentiated source part ($I_i^{\rm S}$) up to 1.5PN order by

$$\mathscr{U}_i^{\rm S} = \dot{I}_i^{\rm S} + \frac{2GM}{\phi_0^{\rm SF}c^3}\int_{-\infty}^{T_{\rm R}} dV \overset{(4)}{I_k^{\rm S}}(V)\left[\ln\left(\frac{T_{\rm R}-V}{2\tau_0}\right)+1\right]. \tag{5.35}$$

5.5.1 *Instantaneous Contribution*

The instantaneous scalar contribution at 1.5PN order is,

$$\mathscr{G}^{\rm inst,S} = \mathscr{G}^{\rm inst,S}_{-1\rm PN} + \mathscr{G}^{\rm inst,S}_{\rm N} + \mathscr{G}^{\rm inst,S}_{0.5\rm PN} + \mathscr{G}^{\rm inst,S}_{1\rm PN} + \mathscr{G}^{\rm inst,S}_{1.5\rm PN}, \tag{5.36}$$

where the Newtonian contribution corresponds to the dominant contribution in GR at 2.5PN order. The results at 1.5PN order in Harmonic coordinates are,

$$\mathscr{G}^{\rm inst,S}_{-1\rm PN} = \frac{G_{AB}^2M^3\nu^2}{c^3r}\dot{\phi}\left(\frac{4}{3}\zeta'\mathscr{S}_-^2\right), \tag{5.37}$$

$$
\begin{aligned}
\mathscr{G}^{\text{inst,S}}_{\text{N}} &= \frac{G^2_{AB}M^3\nu^2}{c^5 r}\dot{\phi}\left(\frac{2}{3}\zeta'\right)\Bigg\{\Bigg[\frac{1}{5}\Big(\left(-53-20\gamma_{AB}+3\nu\right)\mathscr{S}_-^2+4\mathscr{S}_+^2 \\
&\quad -2\mathscr{S}_-\mathscr{S}_+\left(3X_{AB}-10\bar{\beta}_-+10X_{AB}\bar{\beta}_+\right)\Big)-4(-1+\gamma_{AB})\langle\theta\rangle\mathscr{S}_-\Bigg]\frac{G_{AB}M}{r} \\
&\quad +\left[\left(\frac{9}{5}+\frac{24}{5}\nu\right)\mathscr{S}_-^2-\frac{18}{5}X_{AB}\mathscr{S}_+\mathscr{S}_- - 12\langle\theta\rangle\mathscr{S}_- \frac{6}{5}\mathscr{S}_+^2\right]\dot{r}^2 \\
&\quad +\left[\frac{6}{5}X_{AB}\mathscr{S}_+\mathscr{S}_- + \left(3+\frac{3}{5}\nu+2\gamma_{AB}-\frac{13}{5}\nu\right)\mathscr{S}_-^2+4\langle\theta\rangle\mathscr{S}_- + \frac{4}{5}\mathscr{S}_+^2\right]\mathrm{v}^2\Bigg\},
\end{aligned}
\tag{5.38}
$$

$$
\mathscr{G}^{\text{inst,S}}_{0.5\text{PN}} = \frac{G^2_{AB}M^3\nu^2}{c^6 r}\dot{\phi}\left(\frac{2\zeta'\mathscr{S}_-^2}{3}\right)\left\{\gamma_{AB}-4\zeta'\left[X_{AB}\mathscr{S}_+-(4\nu-1)\mathscr{S}_-\right]\mathscr{S}_-\right\}\frac{G_{AB}M}{r}\dot{r}, \tag{5.39}
$$

$$
\begin{aligned}
\mathscr{G}^{\text{inst,S}}_{1\text{PN}} &= \frac{G^2_{AB}M^3\nu^2}{c^7 r}\dot{\phi}\left(\zeta'\right)\left\{\mathscr{G}^{\text{r}^2}_{1\text{PN}}\left(\frac{G_{AB}M}{r}\right)^2+\mathscr{G}^{\text{r,v}^2}_{1\text{PN}}\left(\frac{G_{AB}M}{r}\right)\mathrm{v}^2+\mathscr{G}^{\text{r},\dot{r}^2}_{1\text{PN}}\left(\frac{G_{AB}M}{r}\right)\dot{r}^2\right. \\
&\quad \left. +\mathscr{G}^{\text{v}^4}_{1\text{PN}}\mathrm{v}^4+\mathscr{G}^{\text{v}^2\dot{r}^2}_{1\text{PN}}\mathrm{v}^2\dot{r}^2+\mathscr{G}^{\dot{r}^4}_{1\text{PN}}\dot{r}^4\right\},
\end{aligned}
\tag{5.40}
$$

where the 1PN coefficients explicitly read

$$
\begin{aligned}
\mathscr{G}^{\text{r}^2}_{1\text{PN}} &= \frac{2}{525}\left(-1171+5134\nu-490\gamma_{AB}+1960\gamma_{AB}\nu-1464\nu^2-350X_{AB}(9+4\gamma_{AB}-6\nu)\bar{\beta}_-\right. \\
&\quad \left.+350X_{AB}\delta_-\right)\mathscr{S}_-^2+\frac{2}{525}\left(280X_{AB}(10+\gamma_{AB}(2-8\nu)+\nu)\theta_-+140(-3+4\gamma_{AB})X_{AB}^2\theta_+\right)\mathscr{S}_- \\
&\quad +\frac{4}{105}\left(28(-5+8\nu)\bar{\beta}_-+X_{AB}(212+77\gamma_{AB}-39\nu+28(5+8\nu)\bar{\beta}_+)\right)\mathscr{S}_+\mathscr{S}_- \\
&\quad +\frac{2}{525}\left[140\left(20\bar{\beta}_-(2\nu-1)+(4\gamma_{AB}-7)X_{AB}\right)\theta_+-140\left(20\bar{\beta}_-X_{AB}+4\gamma_{AB}+14\nu-5\right)\theta_-\right]\mathscr{S}_+ \\
&\quad +\frac{2}{105}\left(-251-98\gamma_{AB}+18\nu\right)\mathscr{S}_+^2+\frac{32}{3}\gamma_{AB}X_{AB}\theta_-\theta_+ + \frac{8}{15\left(\mathscr{S}_-^2-\mathscr{S}_+^2\right)^2}\left[-2X_{AB}\theta_-\mathscr{S}_-^5\right. \\
&\quad +2\theta_+\left(X_{AB}\mathscr{S}_++20\gamma_{AB}(1-2\nu)\theta_+\right)\mathscr{S}_+^4+\left(10\gamma_{AB}(-1+10\nu)\theta_-^2-21\theta_-\mathscr{S}_++2X_{AB}\mathscr{S}_+\theta_+\right)\mathscr{S}_-^4 \\
&\quad +\left(10\gamma_{AB}X_{AB}\theta_-^2+\theta_+(21\mathscr{S}_+-10\gamma_{AB}X_{AB}\theta_+)+4\theta_-(X_{AB}\mathscr{S}_++5\gamma_{AB}(1-10\nu)\theta_+)\right)\mathscr{S}_-^3\mathscr{S}_+ \\
&\quad -\left(10\gamma_{AB}X_{AB}\theta_-^2+\theta_+(21\mathscr{S}_+-10\gamma_{AB}X_{AB}\theta_+)+2\theta_-(X_{AB}\mathscr{S}_++20\gamma_{AB}(1-2\nu)\theta_+)\right)\mathscr{S}_-\mathscr{S}_+^3 \\
&\quad \left.+\left(20\gamma_{AB}(1-2\nu)\theta_-^2+21\theta_-\mathscr{S}_++2\theta_+(-2X_{AB}\mathscr{S}_++5\gamma_{AB}(-1+10\nu)\theta_+)\right)\mathscr{S}_-^2\mathscr{S}_+^2\right] \\
&\quad +\frac{\alpha^2_{AB}}{(1-\zeta')^2}\left\{\frac{\mathscr{S}_-}{8400}\left[\mathscr{S}_-\left(169392+7892(2+\gamma_{AB})^2\nu^2-16\left(-18707-7630\bar{\beta}_++700\bar{\beta}_+^2\right.\right.\right.\right. \\
&\quad -700\delta_+-350\langle\varepsilon\rangle\big)\gamma_{AB}+4\left(49367+17500\bar{\beta}_++700\delta_++350\langle\varepsilon\rangle\right)\gamma^2_{AB}+560\left(103+20\bar{\beta}_+\right)\gamma^3_{AB} \\
&\quad +6300\gamma^4_{AB}+6720\bar{\beta}_+-44800\bar{\beta}_+^2+11200\delta_++5600\langle\varepsilon\rangle+\big(-63808+5600(2+\gamma_{AB})^2\bar{\beta}_+ \\
&\quad -110848\gamma_{AB}-79092\gamma^2_{AB}-27860\gamma^3_{AB}-4025\gamma^4_{AB}-67200(4+\gamma_{AB})\bar{\beta}_+^2-42000(\delta_-^2-\delta_+^2) \\
&\quad \left.-2800(2+\gamma_{AB})^2(2\delta_++\varepsilon_+)\big)\nu\big)-94080\theta_+\right]-\frac{8\mathscr{S}_-}{15\left(\mathscr{S}_-^2-\mathscr{S}_+^2\right)^2}\left[-21\mathscr{S}_+^4\theta_+\right. \\
&\quad +\theta_-\left(10\gamma_{AB}(1+6\nu)\theta_-+21\mathscr{S}_+\right)\mathscr{S}_-\mathscr{S}_+^2+\theta_+\left(-20\gamma_{AB}(1+6\nu)\theta_-+21\mathscr{S}_+\right)\mathscr{S}_-^2\mathscr{S}_+ \\
&\quad \left.\left.+\left(-21\theta_-\mathscr{S}_++10\gamma_{AB}(1+6\nu)\theta_+^2\right)\mathscr{S}_-^3\right]\right\}, \\
\mathscr{G}^{\text{r},\dot{r}^2}_{1\text{PN}} &= \frac{32}{3}X_{AB}\theta_-\theta_+-\frac{2}{105}\left[\left(-170+897\nu-428\nu^2+\gamma_{AB}(-28+112\nu)+X_{AB}(56(-4+3\nu)\bar{\beta}_-\right.\right. \\
&\quad \left.+140\delta_-)\right)\mathscr{S}_-^2+\left(28(-13+4\gamma_{AB})X^2_{AB}\langle\theta\rangle+14X_{AB}(155-213)\theta_-\right)\mathscr{S}_-+\big(-112(-7+24\nu)\bar{\beta}_- \\
&\quad +X_{AB}(249+804\nu+413\gamma_{AB}+112(-7+4\nu)\bar{\beta}_+)\big)\mathscr{S}_-\mathscr{S}_++(-184+89\nu-28\gamma_{AB})\mathscr{S}_+^2 \\
&\quad \left.+\left(28(-5+4\gamma_{AB})X_{AB}\langle\theta\rangle-1120X_{AB}\langle\Theta\rangle+112\theta_-+280\nu\theta_--448\nu\gamma_{AB}\theta_-+2240\nu\Theta_-\right)\mathscr{S}_+\right] \\
&\quad +\frac{1}{3}\left[\frac{64(-1+2\nu)\mathscr{S}_+\theta_+\left(\mathscr{S}_-\theta_--\mathscr{S}_+\theta_+\right)}{\mathscr{S}_-^2-\mathscr{S}_+^2}\right]-\frac{4}{15\left(\mathscr{S}_-^2-\mathscr{S}_+^2\right)^2}\left[8X_{AB}\theta_-\mathscr{S}_-^5\right. \\
&\quad +\left((20-720\nu)\theta_-^2-8X_{AB}\langle\theta\rangle\mathscr{S}_++19(-9+7\nu)\theta_-\mathscr{S}_+\right)\mathscr{S}_-^4+\big(-80(-1+2\nu)X_{AB}\theta_-^2 \\
&\quad +X_{AB}(-155+101\nu)\theta_-\mathscr{S}_+-80((-1+2\nu)\langle\Theta\rangle+2X_{AB}\nu\Theta_-)\mathscr{S}_++\langle\theta\rangle(-80(-1+2\nu)\theta_-
\end{aligned}
$$

$$
\begin{aligned}
&+(-163+101\nu)\mathcal{S}_+))\,\mathcal{S}_-\mathcal{S}_+^3+\big(20(-3+8\nu)\theta_-^2+(163-101\nu)\theta_-\mathcal{S}_++80(1-2\nu)\Theta_-\mathcal{S}_+\\
&+16X_{AB}\mathcal{S}_+\theta_++40\theta_+^2-800\nu\theta_+^2-80X_{AB}\mathcal{S}_+\Theta_+\big)\,\mathcal{S}_+^2\mathcal{S}_-^2+4\left((5-20\nu)\theta_-^2\right.\\
&+20(-1+2\nu)\Theta_-\mathcal{S}_+-2X_{AB}\mathcal{S}_+\theta_+-15\theta_+^2+40\nu\theta_+^2+20X_{AB}\mathcal{S}_+\Theta_+\big)\,\mathcal{S}_+^4\\
&+\big(40(-1+36\nu)\langle\theta\rangle\theta_-+40X_{AB}(-1+36\nu)\theta_-^2-16X_{AB}\theta_-\mathcal{S}_++80\Theta_-X_{AB}\mathcal{S}_+\\
&+(163-101\nu)\mathcal{S}_+\theta_++80(-1+2\nu)\Theta_+\mathcal{S}_+\big)\,\mathcal{S}_-^3\mathcal{S}_+\big]\\
&-\frac{\alpha_{AB}^2}{120\,(1-\zeta')^2}\,\big\{\big[980+\nu\left(-1172+\gamma_{AB}(-1780+1704\bar\beta_++320\varepsilon_+)+\gamma_{AB}^2(-421+224\bar\beta_+)\right.\\
&+328\gamma_{AB}^3+120\gamma_{AB}^4+1920(\delta_-^2-\delta_+^2)+4128\bar\beta_++5120\bar\beta_+^2+1280\varepsilon_+\big)\\
&+\nu^2\left(212+212\gamma_{AB}-187\gamma_{AB}^2-240\gamma_{AB}^3-60\gamma_{AB}^4+960(\delta_+^2-\delta_-^2)\right)+\gamma_{AB}\left(1364-792\bar\beta_+\right.\\
&+320\delta_+-160\langle\varepsilon\rangle\big)+\gamma_{AB}^2\left(709+128\bar\beta_++80\delta_+\right)+176\gamma_{AB}^3+20\gamma_{AB}^4-4704\bar\beta_++160\bar\beta_+^2\\
&+320\delta_+-640\langle\varepsilon\rangle\big]\,\mathcal{S}_-^2+\big[32(-163+101\nu)\langle\theta\rangle-2560(-1+2\nu)\langle\Theta\rangle-5216X_{AB}\theta_-\\
&+3232X_{AB}\nu\theta_--5120X_{AB}\nu\Theta_-\big]\,\mathcal{S}_-+\big[2560X_{AB}\langle\Theta\rangle-5120\nu\Theta_-\big]\,\mathcal{S}_+\\
&-\frac{32}{\left(\mathcal{S}_-^2-\mathcal{S}_+^2\right)^2}\,\big[(-80(-1+2\nu)\langle\Theta\rangle+X_{AB}(-163+101\nu)\theta_--160X_{AB}\nu\Theta_-)\,\mathcal{S}_-\mathcal{S}_+^4\\
&+\langle\theta\rangle\left(40(1+\gamma_{AB})(1+32\nu)\mathcal{S}_-^2\theta_-+(-163+101\nu)\mathcal{S}_+^3\right)\mathcal{S}_-\mathcal{S}_++80\left((-1+2\nu)\Theta_-\right.\\
&+X_{AB}\Theta_+\big)\,\mathcal{S}_+^5+\left((-163+101\nu)\theta_-\mathcal{S}_+-20(1+\gamma_{AB})(1+32\nu)\theta_+^2\right)\mathcal{S}_-^4\\
&+\left(20(1+\gamma_{AB})(1+32\nu)\theta_-^2+(-163+101\nu)\theta_-\mathcal{S}_++80\mathcal{S}_+((-1+2\nu)\Theta_-+X_{AB}\Theta_+)\right)\mathcal{S}_-^2\mathcal{S}_+^2\\
&+\big(40X_{AB}(1+\gamma_{AB})(1+32\nu)\theta_-^2+\mathcal{S}_+(80X_{AB}\Theta_-+(163-101\nu)\theta_+\\
&+80(-1+2\nu)\Theta_+)\big)\,\mathcal{S}_-^3\mathcal{S}_+\big]\big\},
\end{aligned}
$$

$$
\begin{aligned}
\mathcal{G}_{\rm 1PN}^{r,v^2}&=\Big[-\frac{229}{70}+\frac{227}{30}\nu-\frac{57}{70}\nu^2+\gamma_{AB}\left(-\frac{32}{15}+\frac{4}{15}\nu\right)+\frac{24}{5}\langle\bar\beta\rangle+\frac{8}{15}\nu\left(3X_{AB}\bar\beta_-+2\bar\beta_+\right)\Big]\,\mathcal{S}_-^2\\
&+\frac{4}{15}\left[\left(-17-47\nu-4\gamma_{AB}X_{AB}^2\right)\langle\theta\rangle-2X_{AB}\left(9+31\nu\right)\theta_--40\nu X_{AB}\Theta_--20\left(-1+2\nu\right)\langle\Theta\rangle\right]\mathcal{S}_-\\
&+\left[X_{AB}\left(\frac{286}{105}+\frac{136}{35}\nu+\frac{14}{5}\gamma_{AB}\right)+\frac{16}{15}\left((7-14\nu)\bar\beta_-+X_{AB}(-7+4\nu)\bar\beta_+\right)\right]\mathcal{S}_-\mathcal{S}_+\\
&+\left[-\frac{8}{15}\left(-9+16\nu+8\nu\gamma_{AB}\right)\theta_-+\frac{16}{15}\left(4+\gamma_{AB}\right)X_{AB}\langle\theta\rangle\right]\mathcal{S}_+-\left[\frac{20}{21}+\frac{76}{35}\nu\right]\mathcal{S}_+^2-\frac{32}{3}X_{AB}\theta_-\theta_+\\
&+\frac{1}{\mathcal{S}_-^2-\mathcal{S}_+^2}\left[\frac{24}{5}X_{AB}\theta_-\left(\mathcal{S}_-^3+\mathcal{S}_+^3\right)+\left(\frac{16}{3}(\theta_-^2+\theta_+^2)+\frac{64}{3}\nu(2\theta_+^2-3\theta_-^2)\right)\mathcal{S}_-^2\right.\\
&+\frac{24}{5}\left(-X_{AB}\langle\theta\rangle+(-1+4\nu)\theta_-\right)\mathcal{S}_-^2\mathcal{S}_+-\frac{24}{5}X_{AB}\theta_-\mathcal{S}_-\mathcal{S}_+^2+\frac{16}{3}\left((-1+4\nu)\theta_-^2-\theta_+^2\right)\mathcal{S}_+^2\\
&\left.+\frac{64}{3}(-1+2\nu)\theta_-\left(\langle\theta\rangle+X_{AB}\theta_--\theta_+\right)\mathcal{S}_-\mathcal{S}_+\right],\\
\mathcal{G}_{\rm 1PN}^{v^4}&=\left[\frac{69}{70}-\frac{1679}{210}\nu+\frac{131}{30}\nu^2+\gamma_{AB}\left(\frac{2}{3}-\frac{58}{15}\nu\right)\right]\mathcal{S}_-^2+\left[\langle\theta\rangle\left(\frac{28}{15}-\frac{164}{5}\nu\right)+\frac{88}{15}\nu\theta_+\right]\mathcal{S}_-\\
&-X_{AB}\left[\frac{16}{15}+\frac{16}{15}\nu+\frac{14}{15}\gamma_{AB}\right]\mathcal{S}_-\mathcal{S}_++\left[-\frac{16}{15}X_{AB}\langle\theta\rangle+\frac{16}{5}\nu\theta_-\right]\mathcal{S}_++\left(\frac{61}{105}+\frac{4}{15}\gamma_{AB}-\frac{116}{105}\nu\right)\mathcal{S}_+^2,\\
\mathcal{G}_{\rm 1PN}^{\dot r^4}&=\left[-\frac{127}{14}+\frac{106}{7}\nu+\frac{12}{7}\nu^2-\gamma_{AB}\left(4-2\nu\right)\right]\mathcal{S}_-^2+\left[\langle\theta\rangle\left(12-40\nu\right)+16\nu\theta_+\right]\mathcal{S}_-\\
&-X_{AB}\left[\frac{48}{7}+\frac{86}{7}\nu+6\gamma_{AB}\right]\mathcal{S}_-\mathcal{S}_++\left[-8X_{AB}\langle\theta\rangle+16\nu\theta_-\right]\mathcal{S}_++\left[-\frac{11}{7}+\frac{24}{7}\nu\right]\mathcal{S}_+^2,\\
\mathcal{G}_{\rm 1PN}^{v^2,\dot r^2}&=\left[\frac{299}{35}-\frac{23}{5}\nu-\frac{302}{35}\nu^2+\gamma_{AB}\left(\frac{18}{5}-\frac{16}{5}\nu\right)\right]\mathcal{S}_-^2+\left[-\frac{136}{5}\nu\theta_++\langle\theta\rangle\left(-\frac{64}{5}+\frac{284}{5}\nu\right)\right]\mathcal{S}_-\\
&+\frac{2}{35}X_{AB}\left[128+297\nu+112\gamma_{AB}\right]\mathcal{S}_-\mathcal{S}_++8\left[X_{AB}\langle\theta\rangle-\frac{12}{5}\nu\theta_-\right]\mathcal{S}_++\left[\frac{8}{7}-\frac{26}{35}\nu\right]\mathcal{S}_+^2.
\end{aligned}
\tag{5.41}
$$

Here the coefficient ζ' in DEF convention is

$$
\zeta'=1-\alpha_{AB}\frac{(2+\gamma_{AB})}{2}. \tag{5.42}
$$

This scalar flux contribution goes to zero in the GR limit where $\alpha_{AB} \to 1$ and $\gamma_{AB} \to 0$ implying $\zeta' \to 0$.

5.5.2 Tail Contribution

Following Sect. 5.4.2 and the computation of the tail contributions to the angular momentum flux in GR [19, 20], we compute the tail contribution to the scalar part of the angular momentum flux in ST theories up to 1.5PN order using the discrete Fourier decomposition of the scalar dipolar and quadrupolar mass moments. The tail contribution to the scalar part of the angular momentum flux starts at 0.5PN order, while the tail terms enter the tensorial part starting from 1.5PN order. The tail integrals up to 1.5PN order in ST theory are given by

$$\begin{aligned}
\mathcal{G}_i^{\mathrm{S,Tail}} &= \frac{G\phi_0^{\mathrm{SF}}(1-\zeta')}{\zeta' c^3}\varepsilon_{ijk} \\
&\times \left\{ \frac{2GM}{3\phi_0^{\mathrm{SF}}c^3} \dot{I}_j^{\mathrm{S}} \int_{-\infty}^{T_{\mathrm{R}}} dV \overset{(4)}{I}{}_k^{\mathrm{S}}(V) \left[\ln\left(\frac{T_{\mathrm{R}}-V}{2\tau_0}\right) + 1 \right] \right. \\
&- \frac{2GM}{3\phi_0^{\mathrm{SF}}c^3} \ddot{I}_j^{\mathrm{S}} \int_{-\infty}^{T_{\mathrm{R}}} dV \overset{(3)}{I}{}_k^{\mathrm{S}}(V) \left[\ln\left(\frac{T_{\mathrm{R}}-V}{2\tau_0}\right) + 1 \right] \\
&+ \frac{2GM}{15\phi_0^{\mathrm{SF}}c^5} \ddot{I}_{jl}^{\mathrm{S}} \int_{-\infty}^{T_{\mathrm{R}}} dV \overset{(5)}{I}{}_{kl}^{\mathrm{S}}(V) \left[\ln\left(\frac{T_{\mathrm{R}}-V}{2\tau_0}\right) + \frac{3}{2} \right] \\
&\left. - \frac{2GM}{15\phi_0^{\mathrm{SF}}c^5} \overset{(3)}{I}{}_{jl}^{\mathrm{S}} \int_{-\infty}^{T_{\mathrm{R}}} dV \overset{(4)}{I}{}_{kl}(V) \left[\ln\left(\frac{T_{\mathrm{R}}-V}{2\tau_0}\right) + \frac{3}{2} \right] \right\}.
\end{aligned} \tag{5.43}$$

As shown in Ref. [24], the scalar dipole moment at Newtonian order in polar coordinates is a periodic function of the mean anomaly ℓ and can be decomposed into a Fourier series, i.e.

$$I_i^{\mathrm{S}}(\ell) = \sum_{p=-\infty}^{\infty} I_i^{\mathrm{S}}(p) e^{\mathrm{i}p\ell} \tag{5.44}$$

where $I_i^{\mathrm{S}}(p)$ are discrete Fourier functions of p and since the moments are real $I_i^{\mathrm{S}}(-p) = I_i^{\mathrm{S}}(p)^*$, where the superscript "$*$" denotes a complex conjugation.

We follow the same method as in Ref. [20] and express the tail terms in Eq. (5.43) in terms of the Fourier functions of the multipole moments, averaged over the mean anomaly ℓ. All of the hereditary contributions, apart from the scalar dipolar mass moment, entering Eq. (5.43) will require only relative Newtonian precision for the leading order computations, hence we can use the simple decomposition of Eq. (5.44). For the beyond Newtonian precision, we will follow the approach used in GR to exploit the doubly periodic nature of the dynamics and express the scalar mass dipolar moment as

$$I_i^{\rm S}(\ell) = \sum_{p=-\infty}^{\infty} \sum_{m=-1}^{1} I_i^{\rm S}(p, m) e^{{\rm i}(p+km)\ell} \tag{5.45}$$

using the general formula of Ref. [20]. Note that we here start from $m = -1$ because this is a dipolar contribution, while the GR equivalent starts from $m = -2$.

5.5.2.1 Relative Newtonian Order

In this section, we present the computation of scalar tail contribution up to 1.5PN order at the relative Newtonian precision. We first separate Eq. (5.43) into the four addends and denote the four terms with subscripts I, II, III and IV.

Let us follow the computation for $(\mathcal{G}_i^{\rm S,Tail})_{\rm I}$ at Newtonian order. We start by taking the average over the mean anomaly ℓ, denoted as $\langle\mathcal{G}_i^{\rm S,Tail}\rangle_{\rm I}$, and get

$$\begin{aligned}\langle\mathcal{G}_i^{\rm S,Tail}\rangle_{\rm I} = &\frac{G^2M(1-\zeta')}{3\pi\zeta' c^6}\varepsilon_{ijk}\int_0^{2\pi} d\ell \dot{I}_j^{\rm S} \\ &\times \int_{-\infty}^{T_{\rm R}} dV I_k^{{\rm S}(4)}(V)\left[\ln\left(\frac{T_{\rm R}-V}{2\tau_0}\right)+1\right].\end{aligned} \tag{5.46}$$

We can now introduce $\tau = T_R - V$. Using the Newtonian Fourier decomposition, Eq. (5.44), and the linear dependence of ℓ on time, so that $\ell(t-\tau) = \ell(t) - n\tau$ (see, e.g., Ref. [24]), we obtain

$$\begin{aligned}\langle\mathcal{G}_i^{\rm S,Tail}\rangle_{\rm I} = &\,{\rm i}\frac{2G^2M(1-\zeta')}{3\zeta' c^6}\varepsilon_{ijk}\sum_{p=1}^{\infty} p^5 n^5 \\ &\times \left\{ I_j^{\rm S}(p) I_k^{\rm S}(p)^* \int_0^{\infty} d\tau e^{{\rm i}np\tau}\left[\ln\left(\frac{\tau}{2\tau_0}\right)+1\right]\right. \\ &\left. - I_k^{\rm S}(p) I_j^{\rm S}(p)^* \int_0^{\infty} d\tau e^{-{\rm i}np\tau}\left[\ln\left(\frac{\tau}{2\tau_0}\right)+1\right]\right\}.\end{aligned} \tag{5.47}$$

To solve the integral in the above equation, we use the closed-form solution of the integral

$$\int_0^{\infty} dt e^{{\rm i}\sigma t}\ln\left(\frac{t}{2t_0}\right) = -\frac{1}{\sigma}\left\{\frac{\pi}{2}{\rm sign}(\sigma) + {\rm i}\left[\ln(2|\sigma|t_0)+\gamma_{\rm E}\right]\right\} \tag{5.48}$$

where sign is the signum function ${\rm sign}(\sigma) = |\sigma|/\sigma$ and $\gamma_{\rm E}$ is Euler's constant. Using this integral, the equation above becomes

$$\langle\mathcal{G}_i^{\rm S,Tail}\rangle_{\rm I} = -{\rm i}\frac{2\pi G^2M(1-\zeta')}{3\zeta' c^6}\varepsilon_{ijk}\sum_{p=1}^{\infty} p^4 n^4 I_j^{\rm S}(p) I_k^{\rm S}(p)^*. \tag{5.49}$$

Similarly, we can solve for the other terms in Eq. (5.43) obtaining

$$\langle \mathcal{G}_i^{\mathrm{S,Tail}} \rangle_{\mathrm{II}} = -\mathrm{i}\frac{2\pi G^2 M(1-\zeta')}{3\zeta' c^6} \varepsilon_{ijk} \sum_{p=1}^{\infty} p^4 n^4 I_j^{\mathrm{S}}(p) I_k^{\mathrm{S}}(p)^*,$$

$$\langle \mathcal{G}_i^{\mathrm{S,Tail}} \rangle_{\mathrm{III}} = -\mathrm{i}\frac{2\pi G^2 M(1-\zeta')}{15\zeta' c^8} \varepsilon_{ijk} \sum_{p=1}^{\infty} p^6 n^6 I_{jl}^{\mathrm{S}}(p) I_{kl}^{\mathrm{S}}(p)^*,$$

$$\langle \mathcal{G}_i^{\mathrm{S,Tail}} \rangle_{\mathrm{IV}} = -\mathrm{i}\frac{2\pi G^2 M(1-\zeta')}{15\zeta' c^8} \varepsilon_{ijk} \sum_{p=1}^{\infty} p^6 n^6 I_{jl}^{\mathrm{S}}(p) I_{kl}^{\mathrm{S}}(p)^*. \tag{5.50}$$

We note that the dipolar contributions at Newtonian order, terms I and II above, give the same contribution. The same holds also for the Newtonian quadrupolar contributions, i.e. $\langle \mathcal{G}_i^{\mathrm{S,Tail}} \rangle_{\mathrm{III}} = \langle \mathcal{G}_i^{\mathrm{S,Tail}} \rangle_{\mathrm{IV}}$.

5.5.2.2 Dipolar Moment at 1st Post-Newtonian Order

The leading order scalar tail contribution of Eq. (5.43) requires I_j^{S} at a relative 1PN order dipolar moment for the scalar flux up to 1.5PN order. In this case, the Fourier decomposition of the dipolar moment is "doubly periodic" in two variables $K\ell = (1+k)\ell$ and ℓ, where ℓ is the mean anomaly and K is the periastron advance per revolution and the decomposition is given in Eq. (5.45). After inserting the Fourier decomposition, the average 1PN order scalar tail contribution reads

$$\langle \mathcal{G}_i^{\mathrm{S,Tail}} \rangle = -\mathrm{i}\frac{2G^2 M(1-\zeta')}{3\zeta' c^6} \varepsilon_{ijk} \sum_{p,p';m,m'} (p+mk)(p'+m'k)^3 [p'-p+(m'-m)k] I_j^{\mathrm{S}}(p,m) I_k^{\mathrm{S}}(p',m')$$
$$\times \langle e^{\mathrm{i}(p+p'+mk+m'k)l} \rangle \int_0^{\infty} d\tau e^{-\mathrm{i}(p'+m'k)n\tau} \left[\ln\left(\frac{\tau}{2\tau_0}\right) + 1 \right], \tag{5.51}$$

where the summation over p and p' ranges from $-\infty$ to ∞, the summation over m and m' ranges from -1 to 1 and the angular brackets denote an average over ℓ. , and integral up to 1PN order reads, The explicit Fourier decomposition up to 1PN order can be obtained following Ref. [20], which we leave for future work.

5.6 Discussion

In this Chapter, we present the angular momentum flux emitted by quasi-circular nonspinning binaries in massless scalar-tensor theories up to 1.5PN order building upon the results of Refs. [12, 25, 26]. The flux in scalar-tensor theories consists of two contributions; scalar contributions and tensorial contributions. In the present work, we first compute the tensorial contribution to the flux up to 1.5PN order. Due to the scalar dipolar radiation in scalar-tensor theories, we also find that the memory effect in the tensorial contribution of ST theories starts at 1PN order less relative to the memory contribution in GR unlike the tail contribution which starts at the same PN order, 1.5PN, as GR. This leading order non-linear memory in ST theories can be understood as the re-radiation of GWs due to the leading order dipolar scalar GWs in contrast to GR where the leading order memory effect is understood as the re-radiation of GWs due to the leading order quadrupolar GWs. To compute the leading order contribution to the non-zero integral of memory effects, we use the Keplerian motion in scalar-tensor theories and obtain a pure zero-frequency component of the memory.

Next, to derive the scalar contribution to the angular momentum flux, we first derive the relation of the scalar angular momentum flux to the radiative scalar multipole moments. Then using this relation we first compute the instantaneous contribution up to 1PN order and then compute the hereditary effects to the scalar part in the angular momentum flux. Since there is a dipolar radiation due to the scalar field, we find that both the instantaneous and the hereditary effects in the scalar part start at 1PN order less relative to the leading order quadrupolar term in GR, i.e. the hereditary contribution starts at 0.5PN order and the instantaneous contribution starts at -1PN order relative to GR. To find the closed form expression for the hereditary contribution, we use the discrete Fourier decomposition of the scalar radiative multipole moments (the Fourier decomposition of dipolar multipole moment is given in Appendix A of Ref. [24] and the Fourier decomposition of quadrupolar multipole moment follows the Fourier decomposition of the GR multipole moment).

As a first consistency test of our results, we check that the angular momentum flux coincides with the angular momentum flux of GR in the GR limit. Second, we study our results under the assumption that the sensitivity (s_I) for the isolated black holes holds for the binary black hole systems as well. For isolated black holes $s_I = 1/2$ (see, Refs. [27]), which in our conventions implies that the parameters $\alpha_I = \beta_I = \beta'_I = \beta''_I = 0$. Inserting this into the results obtained for scalar-tensor theories, we observe that as expected our results are indistinguishable from the results of GR for binary black hole systems, i.e. the angular momentum flux matches with the angular momentum flux for GR in Refs. [17–20].

These results should be considered as a step forward to construct waveform templates for *eccentric* orbits in scalar-tensor theories. In the next chapter, we will use the results obtained here for the angular momentum flux and expressions for the energy flux from Ref. [12] to obtain the radiation-reaction force in the EOB formalism for eccentric orbits.

References

1. Christodoulou D (1991) Nonlinear nature of gravitation and gravitational-wave experiments. Phys Rev Lett 67:1486–1489. https://doi.org/10.1103/PhysRevLett.67.1486
2. Wiseman AG, Will CM (1991) Christodoulou's nonlinear gravitational-wave memory: evaluation in the quadrupole approximation. Phys Rev D 44:R2945–R2949. https://doi.org/10.1103/PhysRevD.44.R2945
3. Thorne KS (1992) Gravitational-wave bursts with memory: the Christodoulou effect. Phys Rev D 45:520–524. https://doi.org/10.1103/PhysRevD.45.520

4. Blanchet L, Damour T (1992) Hereditary effects in gravitational radiation. Phys Rev D 46:4304–4319. https://doi.org/10.1103/PhysRevD.46.4304
5. Jain T, Rettegno P (2024) Angular momentum flux in scalar-tensor theories up to 1.5 post-newtonian order. arXiv:2407.10908v1
6. Freire PC, Wex N, Esposito-Farèse G, Verbiest JP, Bailes M, Jacoby BA, Kramer M, Stairs IH, Antoniadis J, Janssen GH (2012) The relativistic pulsar-white dwarf binary PSR J1738+0333 II. The most stringent test of scalar-tensor gravity. Mon Not Roy Astron Soc 423:3328. arXiv:1205.1450. https://doi.org/10.1111/j.1365-2966.2012.21253.x [astro-ph.GA]
7. Will CM (2014) The confrontation between general relativity and experiment. Living Rev Rel 17(4). arXiv:1403.7377. https://doi.org/10.12942/lrr-2014-4 [gr-qc]
8. Lang RN (2014) Compact binary systems in scalar-tensor gravity. II. Tensor gravitational waves to second post-Newtonian order. Phys Rev D 89:084014. arXiv:1310.3320. https://doi.org/10.1103/PhysRevD.89.084014 [gr-qc]
9. Lang RN (2015) Compact binary systems in scalar-tensor gravity. III. Scalar waves and energy flux. Phys Rev D 91:084027. https://doi.org/10.1103/PhysRevD.91.084027. https://arxiv.org/abs/1411.3073 arXiv:1411.3073 [gr-qc]
10. Bernard L (2018) Dynamics of compact binary systems in scalar-tensor theories: equations of motion to the third post-Newtonian order. Phys Rev D 98:044004. https://doi.org/10.1103/PhysRevD.98.044004. arXiv:1802.10201 [gr-qc]
11. Bernard L (2019) Dynamics of compact binary systems in scalar-tensor theories: II. Center-of-mass and conserved quantities to 3PN order. Phys Rev D 99:044047. https://doi.org/10.1103/PhysRevD.99.044047. arXiv:1812.04169 [gr-qc]
12. Bernard L, Blanchet L, Trestini D (2022) Gravitational waves in scalar-tensor theory to one-and-a-half post-Newtonian order. JCAP 08(08):008. https://doi.org/10.1088/1475-7516/2022/08/008. arXiv:2201.10924 [gr-qc]
13. Damour T, Esposito-Farese G (1992) Tensor multiscalar theories of gravitation. Class Quant Grav 9:2093–2176. https://doi.org/10.1088/0264-9381/9/9/015
14. Damour T, Esposito-Farese G (1993) Nonperturbative strong field effects in tensor - scalar theories of gravitation. Phys Rev Lett 70:2220–2223. https://doi.org/10.1103/PhysRevLett.70.2220
15. Damour T, Esposito-Farese G (1996) Testing gravity to second postNewtonian order: a field theory approach. Phys Rev D 53:5541–5578. https://doi.org/10.1103/PhysRevD.53.5541. arXiv:gr-qc/9506063
16. Thorne KS (1980) Multipole expansions of gravitational radiation. Rev Mod Phys 52:299–339
17. Peters PC (1964) Gravitational radiation and the motion of two point masses. Phys Rev 136:B1224–B1232. https://doi.org/10.1103/PhysRev.136.B1224
18. Junker W, Schäfer G (1992) Binary systems: higher order gravitational radiation damping and wave emission. Mon Not Roy Astron Soc 254:146–164
19. Gopakumar A, Iyer BR (1997) Gravitational waves from inspiralling compact binaries: angular momentum flux, evolution of the orbital elements and the wave form to the second postNewtonian order. Phys Rev D 56:7708–7731. https://doi.org/10.1103/PhysRevD.56.7708, arXiv:gr-qc/9710075
20. Arun KG, Blanchet L, Iyer BR, Sinha S (2009) Third post-Newtonian angular momentum flux and the secular evolution of orbital elements for inspiralling compact binaries in quasi-elliptical orbits. Phys Rev D 80:124018. https://doi.org/10.1103/PhysRevD.80.124018. arXiv:0908.3854 [gr-qc]
21. Rieth R, Schaefer G (1997) Spin and tail effects in the gravitational-wave emission of compact binaries. Class Quant Grav 14:2357–2380. https://doi.org/10.1088/0264-9381/14/8/029
22. Racine E, Buonanno A, Kidder LE (2009) Recoil velocity at 2PN order for spinning black hole binaries. Phys Rev D 80:044010. https://doi.org/10.1103/PhysRevD.80.044010, arXiv:0812.4413 [gr-qc]
23. Faye G, Marsat S, Blanchet L, Iyer BR (2012) The third and a half post-Newtonian gravitational wave quadrupole mode for quasi-circular inspiralling compact binaries. Class Quant Grav 29:175004. https://doi.org/10.1088/0264-9381/29/17/175004, arXiv:1204.1043 [gr-qc]

24. Jain T (2023) Nonlocal-in-time effective one body Hamiltonian in scalar-tensor gravity at third post-Newtonian order. Phys Rev D 107:084018. https://doi.org/10.1103/PhysRevD.107.084018, arXiv:2301.01070 [gr-qc]
25. Lange J, O'Shaughnessy R, Rizzo M (2018) Rapid and accurate parameter inference for coalescing, precessing compact binaries. arXiv:1805.10457 [gr-qc]
26. Lange J et al (2017) Parameter estimation method that directly compares gravitational wave observations to numerical relativity. Phys Rev D 96:104041. https://doi.org/10.1103/PhysRevD.96.104041, arXiv:1705.09833 [gr-qc]
27. Hawking S, Hartle J (1972) Energy and angular momentum flow into a black hole. Commun Math Phys 27:283–290. https://doi.org/10.1007/BF01645515

Chapter 6
Radiation Reaction Force for Scalar-Tensor Theories in Effective-One-Body Formalism

We derive the radiation reaction force for an inspiralling nonspinning binary system of compact objects in massless scalar-tensor theories within the effective-one-body formalism. Here, we compute the radiation reaction force for *eccentric* orbits up to 1PN order. These results then be implemented in the effective-one-body based waveform models to construct waveform templates for generic orbit binaries in scalar-tensor theories of gravity.

This is based on the work in preparation in Ref. [1].

6.1 Introduction

The effective-one-body (EOB) Hamiltonian derived in the first two chapters of this thesis and Refs. [2–4] describe the conservative dynamics of general binary orbits (quasicircular, elliptic-like, or hyperbolic-like) in scalar-tensor (ST) theories up to 3PN order. However, the currently existing implementations of the radiation reaction force [5] and the emitted waveforms for scalar-tensor theories in the EOB-based waveform models `TEOBResumS` and `SEOBNR` are limited to the case of quasicircular, inspiralling orbits up to 1PN order.

Recently, numerical relativity simulations have explored the dynamics (and hence, the waveform) of binary systems in configurations with eccentric orbits [6, 7] and high-velocity encounters of black holes [8, 9] in GR. We anticipate that such simulations of general orbits will also be performed in scalar-tensor theories in the future. With this motivation, we derive the radiation reaction force along eccentric orbits within the EOB formalism for scalar-tensor theories. (We leave a corresponding generalisation of the EOB gravitational waveforms to future work.)

The two-body system's energy and angular momentum flux are required to derive the radiation reaction force for eccentric orbits in the EOB formalism. The energy flux for non-spinning inspiralling compact objects in scalar-tensor theories is derived

T. Jain, *Surfing the Gravitational Waves*, Springer Theses,
https://doi.org/10.1007/978-3-031-96180-9_6

in Ref. [10] up to 1.5 PN order for generic orbits and the angular momentum flux for generic orbits is computed in Chap. 5 up to 1 PN order.

This chapter is organised as follows. In Sect. 6.2, we briefly recall the scalar-tensor parameter conventions and then describe the radiation reaction approach in Sect. 6.3. This is followed by the derivation of the coordinate transformation in Sect. 6.4 required to map the energy and angular momentum flux derived in Ref. [10] and Chapter 5 into EOB coordinates and finally the derivation of the radiation reaction force in scalar-tensor theories in Sect. 6.5.

6.2 Scalar-Tensor Parameters

The massless scalar-tensor theories considered in this Chapter are described by the coupling of the usual Einstein-Hilbert action to the scalar field φ according to

$$S = \frac{c^4}{16\pi G}\int d^4x\sqrt{-g}(R - 2g^{\mu\nu}\partial_\mu\varphi\partial_\nu\varphi) + S_m[\Psi, \mathscr{A}(\varphi)^2 g_{\mu\nu}] \,. \tag{6.1}$$

Here, $g_{\mu\nu}$ is the Einstein metric, R is the Ricci scalar, Ψ collectively denotes the matter fields, $g \equiv \det(g_{\mu\nu})$ and G is the Newtonian constant.

Before proceeding to the computations, we recall that in this chapter we use the conventions and notations of Refs. [11, 12] for binary parameters describing the scalar field effects on the dynamics, i.e., DEF conventions for the ST parameters. Therefore, the energy flux derived in Ref. [10] in Jordan frame parameters is converted to the dimensionless Einstein frame parameters following the conversion relation given in Table 2.1. We also use geometric units with $c = G = 1$ unless specified explicitly.

6.3 Overview of Radiation Reaction Force

In this chapter, we will follow the *balance* approach, i.e. we use the balance between the energy and the angular momentum losses due to the gravitational radiation at infinity, implemented within the EOB formalism in Refs. [13, 14] for GR to derive the radiation reaction force in scalar-tensor theories. This EOB formalism-based balance approach is based on the balance approach developed by Refs. [15–17] in the PN formalism.

When considering the motion within the plane, the equations of motion (EOM) in the Hamiltonian formalism after taking into account the radiation reaction force due to the energy and angular momentum losses read,

$$\dot{r} = \frac{\partial H}{\partial p_r}, \quad \dot{p}_r = -\frac{\partial H}{\partial r} + \mathscr{F}_r \,,$$

$$\dot{\phi} = \frac{\partial H}{\partial p_\phi} , \quad \dot{p}_\phi = -\frac{\partial H}{\partial \phi} + \mathscr{F}_\phi . \tag{6.2}$$

where we use polar coordinates (r, ϕ) with their conjugate momenta (p_r, p_ϕ) to describe the motion of the particles, $\mathscr{F}_\phi$ and $\mathscr{F}_r$ to denote the independent components of the radiation reaction force, H denotes the Hamiltonian, and a 'dot' denotes differentiation with respect to time.

The force components $\mathscr{F}_\phi$ and $\mathscr{F}_r$ can be determined for the energy

$$E_{\text{system}} = H(r, p_r, p_\phi) - (m_A + m_B)c^2 , \tag{6.3}$$

and the angular momentum

$$J_{\text{system}} = p_\phi , \tag{6.4}$$

of the binary system with masses m_A and m_B by balancing the energy and angular momentum losses derived using the above EOM with the corresponding energy Φ_E and angular momentum Φ_J fluxes emitted in the form of gravitational waves at infinity. However, we can not simply equate the energy and angular momentum losses of the system to $-\Phi_E$ and $-\Phi_J$, respectively, as the energy and angular momentum of the system have additional terms called the Schott terms say E_{Schott} and J_{Schott} (see, Ref. [18]) due to the system's interaction with the radiation field.

Using Hamilton's EOM and the fact that H is independent of ϕ we obtain that the time derivatives of the energy and angular momentum (radiation losses) of the system are,

$$\begin{aligned} \dot{E}_{\text{system}} &= \mathscr{F}_r \dot{r} + \mathscr{F}_\phi \dot{\phi} , \\ \dot{J}_{\text{system}} &= \mathscr{F}_\phi . \end{aligned} \tag{6.5}$$

The balance equation after including the Schott terms read

$$\begin{aligned} \dot{E}_{\text{system}} + \dot{E}_{\text{Schott}} + \Phi_E &= 0 , \\ \dot{J}_{\text{system}} + \dot{J}_{\text{Schott}} + \Phi_J &= 0 . \end{aligned} \tag{6.6}$$

Now, on inserting Eq. (6.5) into the balance equations, the equations for the two components of the force read,

$$\begin{aligned} \mathscr{F}_r \dot{r} &= -\mathscr{F}_\phi \dot{\phi} - \dot{E}_{\text{Schott}} - \Phi_E , \\ \mathscr{F}_\phi &= -\dot{J}_{\text{Schott}} - \Phi_J . \end{aligned} \tag{6.7}$$

The far zone energy and angular momentum fluxes are gauge invariant but the radiation reaction forces $\mathscr{F}_r$ and $\mathscr{F}_\phi$, and the Schott terms are gauge dependent quantities. The gauge freedom in radiation reaction force at 2.5PN in GR was first discussed by Refs. [15, 16] where they found that this gauge freedom is related to the freedom to change the coordinate. Following this, Ref. [13] showed that the gauge freedom of radiation reaction term is related to gauge freedom in defining Schott terms, i.e., we can consider a vanishing J_{Schott}.

The relations between the fluxes and the radiation reaction forces based on the circular-orbit relation $\Phi_E^{\text{qc}} = \Omega\Phi_J^{\text{qc}}$ are,

$$\begin{aligned}\mathscr{F}_\phi^{\text{qc}} &= -\frac{\Phi_E^{\text{qc}}}{\Omega},\\ \mathscr{F}_r^{\text{qc}} &= \mathscr{F}_\phi^{\text{qc}}\frac{p_r}{p_\phi} = -\frac{\Phi_E^{\text{qc}} p_r}{\Omega p_\phi},\end{aligned} \tag{6.8}$$

where Ω is the orbital frequency. These relations used for evolving the Hamilton's equations are not valid for generic orbits and do not include Schott terms. The aim of this work is to extend the quasi-circular orbit analysis to the eccentric orbits such that the radiation reaction force can be factorised as

$$\begin{aligned}\mathscr{F}_r &= \mathscr{F}_r^{\text{qc}}\mathscr{F}_r^{\text{ecc}},\\ \mathscr{F}_\phi &= \mathscr{F}_\phi^{\text{qc}}\mathscr{F}_\phi^{\text{ecc}},\end{aligned} \tag{6.9}$$

where $\mathscr{F}^{qc}$ and $\mathscr{F}^{\text{ecc}}$ are the quasi-circular and the eccentric contributions to the radiation reaction force, respectively. Therefore, to derive the radiation reaction force in the EOB coordinates using the above balance equations, we use the energy flux and angular momentum flux in the EOB coordinates (see, Sect. 6.5.1 for their explicit expressions) and write a generic ansatz with the unknown coefficients for the Schott terms. Then, we calculate these unknown coefficients in the Schott terms by requiring that the force be reduced to the radiation reaction force in the limit of quasi-circular orbits. The unknown coefficients entering in the general ansatz are specified by demanding that the radiation reaction force given in Eq. (6.7) in the quasi-circular limit reduces to the following conditions,

$$\begin{aligned}\mathscr{F}_\phi &= -\Phi_J + \mathscr{O}\left(p_r^2\right) + \mathscr{O}\left(\dot{p}_r\right),\\ \frac{\mathscr{F}_r p_\phi}{\mathscr{F}_\phi p_r} &= 1 + \mathscr{O}\left(p_r^2\right),\end{aligned} \tag{6.10}$$

where the radial momenta p_r and its first derivative $\dot{p}_r$ are zero for circular orbits.

6.4 Coordinate Transformations

In Chap. 5 and Ref. [10, 19], the angular momentum and the energy flux for scalar-tensor theory are derived in a harmonic coordinate system. As this chapter aims to derive the radiation reaction within the EOB formalism, we transform the fluxes into the EOB coordinate system. To do so, we use (i) the transformation between ordinary (hereafter, ADM-like as the ADM coordinates fall under the special category of coordinate system in which the Lagrangian is not dependent on the acceleration and its higher derivatives (see, Ref. [20]).)[1] and EOB coordinates derived in Refs. [2, 3, 21] and (ii) the transformation between ADM-like and harmonic coordinates derived in Refs. [3, 21] to derive the transformation from the harmonic to EOB coordinates.

Here, we first define the family of fundamental scalars associated with each coordinate system as

$$v_h^2 = (\mathbf{v}_h \cdot \mathbf{v}_h),\quad \dot{r}_h^2 = (\mathbf{n}_h \cdot \mathbf{v}_h)^2,\quad \frac{1}{r_h}\,, \tag{6.11}$$

$$\begin{aligned} v_a^2 &= (\mathbf{v}_a \cdot \mathbf{v}_a),\quad p_a^2 = (\mathbf{p}_a \cdot \mathbf{p}_a),\quad \dot{r}_a^2 = (\mathbf{n}_a \cdot \mathbf{v}_a)^2,\\ p_{r,a}^2 &= (\mathbf{n}_a \cdot \mathbf{p}_a)^2,\quad \frac{1}{r_a}\,, \end{aligned} \tag{6.12}$$

$$p_e^2 = (\mathbf{p}_e \cdot \mathbf{p}_e),\quad p_{r,e}^2 = (\mathbf{n}_e \cdot \mathbf{p}_e)^2,\quad \frac{1}{r_e}\,, \tag{6.13}$$

where the subscripts 'h', 'a' and 'e' signify the harmonic coordinate system, ADM-like coordinate system and the EOB coordinate system, respectively. The boldfaced letters are used to denote a vector quantity. Note that, in this Chapter, we use a '*tilde*' to define dimensionless variables, i.e.,

$$\tilde{\mathbf{p}}_a = \frac{\mathbf{p}_a}{\mu}\,,\qquad \tilde{p}_{r,a}^2 = (\frac{p_{r,a}}{\mu})^2\,, \tag{6.14}$$

$$\tilde{\mathbf{p}}_e = \frac{\mathbf{p}_e}{\mu}\,,\qquad \tilde{p}_{r,e}^2 = (\frac{p_{r,e}}{\mu})^2\,, \tag{6.15}$$

$$\tilde{\mathbf{p}}_h = \frac{\mathbf{p}_h}{\mu}\,,\qquad \tilde{p}_{r,h}^2 = (\frac{p_{r,h}}{\mu})^2\,. \tag{6.16}$$

ADM to harmonic As mentioned in Chap. 2, the acceleration dependence in the Lagrangian starts at 2PN order, so that the transformation between the ADM-like and the harmonic coordinate system also starts at 2PN order. The explicit 2PN coordinate transformation is derived in Ref. [3] and the explicit 3PN coordinate transformation is derived in Ref. [21]. The transformation up to 2PN order is,

$$\mathbf{r}_h = \mathbf{r}_a + \mathcal{O}\left(\frac{1}{c^4}\right)\,, \tag{6.17}$$

$$\mathbf{v}_h = \frac{d\mathbf{r}_h}{dt} = \mathbf{v}_a + \mathcal{O}\left(\frac{1}{c^4}\right)\,, \tag{6.18}$$

$$\mathbf{p}_a = M\nu\left\{\mathbf{v}_a + \frac{1}{c^2}\left(\nu\frac{G_{\text{AB}}M}{r_a}\dot{r}_a\,\mathbf{n}_a + v_a^2\frac{(1-3\nu)}{2}\mathbf{v}_a + (3+\nu+2\gamma_{\text{AB}})\frac{G_{\text{AB}}M}{r_a}\mathbf{v}_a\right)\right\}$$

[1] Here, by an ordinary coordinate systems, we mean that the equations of motion of the system are only dependent on the positions and velocities of the two bodies.

$$+\mathcal{O}\left(\frac{1}{c^4}\right), \tag{6.19}$$

$$\mathbf{v}_a = \tilde{\mathbf{p}}_a + \frac{1}{c^2}\left(-\nu\frac{G_{\mathrm{AB}}M}{r_a}\tilde{p}_{r,a}\,\mathbf{n}_a - \tilde{p}_a^2\frac{(1-3\nu)}{2}\tilde{\mathbf{p}}_a - (3+\nu+2\gamma_{\mathrm{AB}})\frac{G_{\mathrm{AB}}M}{r_a}\tilde{\mathbf{p}}_a\right)$$
$$+\mathcal{O}\left(\frac{1}{c^4}\right), \tag{6.20}$$

$$\mathbf{n}_h = \mathbf{n}_a + \mathcal{O}\left(\frac{1}{c^4}\right). \tag{6.21}$$

Therefore, the explicit transformation of the fundamental scalar quantities from the harmonic coordinate system to the ADM-like coordinate system is

$$\begin{aligned}\mathrm{v}_h^2 &= \mathrm{v}_a^2 + \mathcal{O}\left(\frac{1}{c^4}\right)\\ &= \tilde{p}_a^2 + \frac{1}{c^2}\left(-2\nu\frac{G_{\mathrm{AB}}M}{r_a}\tilde{p}_{r,a}^2 - (1-3\nu)\tilde{p}_a^4 - 2(3+\nu+2\gamma_{\mathrm{AB}})\frac{G_{\mathrm{AB}}M}{r_a}\tilde{p}_a^2\right) + \mathcal{O}\left(\frac{1}{c^4}\right),\end{aligned} \tag{6.22}$$

$$\begin{aligned}\dot{r}_h^2 &= \dot{r}_a^2 + \mathcal{O}\left(\frac{1}{c^4}\right)\\ &= \tilde{p}_{r,a}^2 + \frac{1}{c^2}\left(-2\nu\frac{G_{\mathrm{AB}}M}{r_a}\tilde{p}_{r,a}^2 - \tilde{p}_a^2(1-3\nu)\tilde{p}_{r,a}^2 - 2(3+\nu+2\gamma_{\mathrm{AB}})\frac{G_{\mathrm{AB}}M}{r_a}\tilde{p}_{r,a}^2\right)\\ &\quad+\mathcal{O}\left(\frac{1}{c^4}\right),\end{aligned} \tag{6.23}$$

$$\frac{1}{r_h} = \frac{1}{r_a} + \mathcal{O}\left(\frac{1}{c^4}\right). \tag{6.24}$$

From the above transformations, we verify that the harmonic to ADM-like coordinate transformation of scalar-tensor theory reduces to the coordinate transformation of GR given in Ref. [13] in the GR limit, i.e. in the limit $\gamma_{AB} \to 0$.

Although we derive the radiation reaction force only up to 1 PN order in this work, we also investigate the coordinate transformation at 1.5PN order. In the above equations, we only compute the transformation involving the conservative part up to the 1PN order. However, due to the scalar dipolar radiation in the scalar-tensor theory, the radiation reaction terms enter at 1.5PN order and there might remain a possible transformation (gauge choice) in the radiation reaction terms. The lowest order, i.e. a 1.5PN order, dissipative relative acceleration term in harmonic coordinates given in Ref. [10] is,

$$\mathbf{a}_h^{1.5\mathrm{PN}} = \frac{4\zeta\mathcal{S}_-^2}{3c^3 r_h}\frac{G_{\mathrm{AB}}^2 M^2\nu}{r_h^2}\left[3\dot{r}_h\mathbf{n}_h - \mathbf{v}_h\right]. \tag{6.25}$$

To investigate the alternative gauge choices for radiation reaction at the leading order, we first write the general form of the radiation reaction term following Refs. [15, 16] for the binary system without spins and tidal effects using the fact that, (i) the term must be a correction to the Newtonian acceleration implying that it should be proportional to $(GM)/r^2$, (ii) should vanish in the test-mass limit, i.e., should be proportional to the symmetric mass-ratio ν, (iii) must be non-linear in Newton's constant as it is related to the emission of gravitational radiation (i.e., should contain an extra factor of $(GM)/r$), and (iv) finally, must be odd in velocities, i.e. dissipative. As we work in scalar-tensor theories, Newton's constant is replaced by the theory's effective gravitational constant G_{AB}. Therefore, the general radiation reaction term at the leading order reads

$$\mathbf{a}_{1.5\text{PN}} = \frac{G_{\text{AB}}^2 M^2}{r^3} \nu \left[A_{3/2} \dot{r} \mathbf{n} + B_{3/2} \mathbf{v} \right] , \tag{6.26}$$

where $\mathbf{v}$, and $\mathbf{n}$ denote the relative velocity and separation vector. The values for the parameters $A_{3/2}$ and $B_{3/2}$ in harmonic coordinates can be read from Eq. (6.25). The goal of our investigation is to determine these two arbitrary parameters in ADM-like coordinates, for which it will be sufficient to investigate the gauge freedom in defining the radiation reaction terms.

For this, we take the post-Newtonian expressions for the energy and angular momentum, $\tilde{E} \equiv E/\mu$ and $\tilde{J} \equiv J/\mu$, and calculate $d\tilde{E}/dt$ and $d\tilde{J}/dt$ using the 1PN equations of motion supplemented with 1.5PN radiation reaction as given in Eq. (6.26). The angular momentum and energy are conserved up to 1 PN order but the presence of dissipative effects at 1.5 PN order leads to non-vanishing values of $d\tilde{E}/dt$ and $d\tilde{J}/dt$. Hence, we have the freedom to add as yet unknown terms to $\tilde{E}$ and $\tilde{J}$ at 1.5PN order. However, the terms should be chosen such that they are proportional to ν (vanish in the test-mass limit), $(G_{AB}M)/r$ (non-linear effects for gravitational waves emission), and odd in $\dot{r}$ (dissipative effects). Therefore, the possible terms for $\tilde{E}$ and $\tilde{J}$ are

$$\tilde{E}^* \equiv \tilde{E}_N + \nu\, \dot{r} \left(\frac{G_{AB}M}{r} \right)^2 \alpha_{3/2} , \tag{6.27}$$

$$\tilde{J}^* \equiv \tilde{J}_N + \nu\, \dot{r} \tilde{J}_N \left(\frac{G_{AB}M}{r} \right) \beta_{3/2} , \tag{6.28}$$

where we use a '*' to distinguish the arbitrary terms entering at 1.5 PN order from the conserved quantities up to 1 PN order, $\tilde{E}_N$ and $\tilde{J}_N = \mathbf{x} \times \mathbf{v}$ are the Newtonian order contribution to the energy and angular momentum, $\alpha_{3/2}$ and $\beta_{3/2}$ are the arbitrary parameters entering at 1.5 PN order. The orbital energy and angular momentum loss therefore reads

$$\frac{d\tilde{E}^*}{dt} = \mathbf{v} \cdot \mathbf{a}_{1.5\text{PN}} + \nu\alpha_{3/2} \left[\ddot{r} \left(\frac{G_{\text{AB}}M}{r} \right)^2 - 2\frac{\dot{r}^2}{r} \left(\frac{G_{\text{AB}}M}{r} \right)^2 \right] , \tag{6.29}$$

$$\frac{d\tilde{J}^*}{dt} = \mathbf{x} \times \mathbf{a}_{1.5\text{PN}} + \nu\beta_{3/2} \tilde{J}_N \left[\frac{G_{\text{AB}}M}{r} \ddot{r} - \frac{G_{\text{AB}}M}{r} \frac{\dot{r}^2}{r} \right] . \tag{6.30}$$

Now using the Newtonian-order equations of motion

$$\ddot{r} = -\frac{G_{\mathrm{AB}}M}{r^2} + \frac{v^2}{r} - \frac{\dot{r}^2}{r}\,, \tag{6.31}$$

and acceleration at 1.5PN order from Eq. (6.26), the explicit expressions for the energy and angular momentum loss read,

$$\frac{d\tilde{E}^*}{dt} = \left(\frac{G_{\mathrm{AB}}M}{r}\right)^2 \frac{\nu}{r}\left[A_{3/2}\dot{r}^2 + B_{3/2}v^2 + \alpha_{3/2}\left(v^2 - \frac{G_{AB}M}{r} - 3\dot{r}^2\right)\right]\,, \tag{6.32}$$

$$\frac{d\tilde{J}^*}{dt} = \frac{G_{\mathrm{AB}}M}{r^2}\frac{\nu\tilde{J}_N}{r}\beta_{3/2}\left(v^2 - \frac{G_{AB}M}{r} - 2\dot{r}^2\right) + \left(\frac{G_{\mathrm{AB}}M}{r}\right)^2\frac{\nu\tilde{J}_N}{r}B_{3/2}\,. \tag{6.33}$$

We now use the assumption of energy and angular momentum balance, i.e. the angular momentum and the energy orbital loss correspond to the far-zone fluxes,

$$\frac{d\tilde{E}^*}{dt} = -\left(\frac{d\tilde{E}}{dt}\right)_{\mathrm{far-zone}}\,, \tag{6.34}$$

$$\frac{d\tilde{J}^*}{dt} = -\left(\frac{d\tilde{J}}{dt}\right)_{\mathrm{far-zone}}\,, \tag{6.35}$$

where the leading order far-zone fluxes are

$$\left(\frac{d\tilde{E}}{dt}\right)_{\mathrm{far-zone}} = \frac{4}{3}\zeta\mathscr{S}_-^2\frac{\nu}{r}\left(\frac{G_{\mathrm{AB}}M}{r}\right)^3\,, \tag{6.36}$$

$$\left(\frac{d\tilde{J}}{dt}\right)_{\mathrm{far-zone}} = \frac{4}{3}\zeta\mathscr{S}_-^2\frac{\nu}{r}\left(\frac{G_{\mathrm{AB}}M}{r}\right)^2\tilde{J}_N\,. \tag{6.37}$$

Comparing term-by-term Eq. (6.32) with Eqs. (6.36) and (6.33) with Eq. (6.37), we see that all four arbitrary parameters $A_{3/2}$, $B_{3.2}$, $\alpha_{3/2}$ and $\beta_{3/2}$ are uniquely determined,

$$A_{3/2} = 4\zeta\mathscr{S}_-^2\,, \quad \alpha_{3/2} = \frac{4}{3}\zeta\mathscr{S}_-^2\,, \tag{6.38}$$

$$B_{3/2} = -\frac{4}{3}\zeta\mathscr{S}_-^2\,, \quad \beta_{3/2} = 0\,. \tag{6.39}$$

The far-zone energy and angular momentum fluxes are gauge invariant and therefore, the unique determination of the arbitrary coefficients implies that there is no residual gauge freedom at 1.5PN order in scalar-tensor theories in contrast to the residual gauge freedom at 2.5PN order in GR (see, Refs. [15, 16]). Furthermore, this implies that there is no freedom to change the positions and velocities at 1.5PN order in scalar-tensor theories, i.e. there is no coordinate transformation between ADM and harmonic coordinates at 1.5PN order.

ADM to EOB In this section, we derive the transformation of the fundamental scalars from ADM coordinates to EOB coordinates up to 1PN order. The ADM to EOB canonical transformation up to 3PN order is derived in Refs. [3, 21]. For this work, we consider transformations only up to 1PN order as both the angular momentum and energy flux considered in this work up to 1PN order. The transformation from ADM to EOB coordinates up to 1PN order is

$$r_a = r_e - \frac{r_e}{c^2}\left\{\left[-\frac{\nu}{2}\tilde{p}_e^2 + \frac{G_{\rm AB}M}{r_e}\left(1+\gamma_{\rm AB}+\frac{\nu}{2}\right)\right] - \nu\tilde{p}_{r,e}^2\right\} + \mathcal{O}\left(\frac{1}{c^4}\right), \tag{6.40}$$

$$p_{r,a} = p_{r,e} - \frac{1}{c^2}\left[\frac{\nu}{2}p_{r,e}\left(2\tilde{p}_{r,e}^2 - \tilde{p}_e^2\right)\right] + \mathcal{O}\left(\frac{1}{c^4}\right), \tag{6.41}$$

and hence, the transformation from the EOB to ADM coordinates up to 1 PN order reads

$$r_e = r_a + \frac{r_a}{c^2}\left\{\left[-\frac{\nu}{2}\tilde{p}_a^2 + \frac{G_{\rm AB}M}{r_a}\left(1+\gamma_{\rm AB}+\frac{\nu}{2}\right)\right] - \nu\tilde{p}_{r,a}^2\right\} + \mathcal{O}\left(\frac{1}{c^4}\right), \tag{6.42}$$

$$p_{r,e} = p_{r,a} + \frac{1}{c^2}\left[\frac{\nu}{2}p_{r,a}\left(2\tilde{p}_{r,a}^2 - \tilde{p}_a^2\right)\right] + \mathcal{O}\left(\frac{1}{c^4}\right). \tag{6.43}$$

The explicit transformation of the fundamental scalar quantities in the ADM coordinate system to the EOB coordinate system reads

$$\frac{1}{r_a} = \frac{1}{r_e} + \frac{1}{c^2}\left\{\frac{1}{r_e}\left[-\frac{\nu}{2}\tilde{p}_e^2 + \frac{G_{\rm AB}M}{r_e}\left(1+\gamma_{\rm AB}+\frac{\nu}{2}\right)\right] - \frac{\nu}{r_e}\tilde{p}_{r,e}^2\right\} + \mathcal{O}\left(\frac{1}{c^4}\right), \tag{6.44}$$

$$p_{r,a}^2 = p_{r,e}^2 - \frac{1}{c^2}\left[\nu p_{r,e}^2\left(2\tilde{p}_{r,e}^2 - \tilde{p}_e^2\right)\right] + \mathcal{O}\left(\frac{1}{c^4}\right), \tag{6.45}$$

$$p_a^2 = p_e^2 + \frac{1}{c^2}\left[-\nu p_e^2\tilde{p}_e^2 + \left(p_e^2 - p_{r,e}^2\right)\left(\frac{G_{\rm AB}M}{r_e}(2+\nu+2\gamma_{\rm AB})\right)\right] + \mathcal{O}\left(\frac{1}{c^4}\right). \tag{6.46}$$

Harmonic to EOB Since we want to transform the harmonic coordinate angular momentum flux and energy flux presented in Refs. [10, 19, 22, 23], we combine the two transformations given above to obtain the transformation from harmonic to the EOB coordinates. The explicit expressions of the transformation up to 1 PN order read

$$\dot{r}_h^2 = \tilde{p}_{r,e}^2 + \frac{1}{c^2}\left[-2\nu\tilde{p}_{r,e}^4 + (4\nu-1)\tilde{p}_e^2\tilde{p}_{r,e}^2 - 2(3+2\nu+2\gamma_{\rm AB})\frac{G_{\rm AB}M}{r_e}\tilde{p}_{r,e}^2\right] + \mathcal{O}\left(\frac{1}{c^4}\right), \tag{6.47}$$

$$v_h^2 = \tilde{p}_e^2 + \frac{1}{c^2}\left\{[-1+2\nu]\,\tilde{p}_e^4 + \frac{G_{\rm AB}M}{r_e}\left[\tilde{p}_{r,e}^2(-2-3\nu-2\gamma_{\rm AB}) - \tilde{p}_e^2(4+\nu+2\gamma_{\rm AB})\right]\right\} + \mathcal{O}\left(\frac{1}{c^4}\right), \tag{6.48}$$

$$\frac{1}{r_h} = \frac{1}{r_e} + \frac{1}{c^2}\left[-\frac{\nu}{2r_e}\tilde{p}_e^2 + \frac{G_{\rm AB}M}{r_e^2}\left(1+\gamma_{\rm AB}+\frac{\nu}{2}\right) - \frac{\nu}{r_e}\tilde{p}_{r,e}^2\right] + \mathcal{O}\left(\frac{1}{c^4}\right). \tag{6.49}$$

All the above transformations, ADM to harmonic, ADM to EOB and harmonic to EOB, match with GR results given in Ref. [13] in GR limit, i.e. in the limit $\gamma_{AB} \to 0$.

6.5 Radiation Reaction Force in Scalar-Tensor Theory

In this section, we first derive the expression for the energy and angular momentum flux in the EOB coordinates, then we write a general ansatz for Schott terms, and finally use this information in Eqs. (6.7) and (6.10) to derive the eccentric orbit radiation reaction force within the EOB formalism for non-spinning binary systems without tides.

Now, let us remember that there are two contributions to the fluxes in scalar-tensor theory, i.e. the tensorial contribution and the scalar contribution. We compare the scalar contribution of the energy flux and the corresponding tensorial contribution of the energy flux in the quasi-circular orbit limit for equal mass binaries with the total flux for scalar-tensor theories in Fig. 6.1 and observe that the scalar contribution is much smaller compared to the tensorial contribution. Therefore, for the sake of simplicity, we only consider the tensorial contribution to the flux and leave for future work the incorporation of the scalar contribution to the radiation reaction force.

6.5.1 *Flux in Effective-One-Body Coordinates*

In this section, we use the coordinate transformation to express the tensorial part of the energy and angular momentum flux given in harmonic coordinates in Ref. [10] and Chap. 5, respectively, in EOB coordinates.

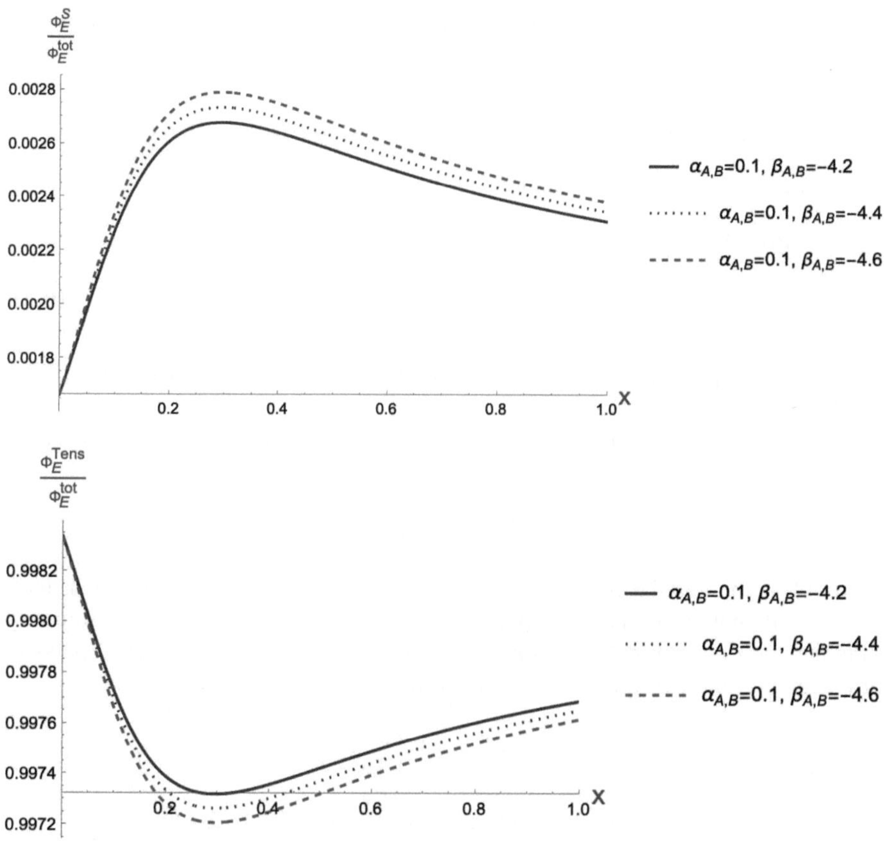

Fig. 6.1 The ratio of a) the scalar contribution of the energy flux Φ_E^S and b) the tensorial contribution of the energy flux Φ_E^{Tens} to the total energy flux contribution $\Phi_E^{tot}(= \Phi_E^S + \Phi_E^{\text{Tens}})$ in the quasi-orbital limit for equal-mass binaries as a function of the frequency-dependent variable $\mathrm{x} \equiv \left(\frac{\mathrm{G_{AB}M}\omega}{\mathrm{c}^3}\right)^{2/3}$ where ω is the orbital frequency of the quasi-circular orbit. The blue, purple and red curves correspond to different values of the scalar-tensor parameter $\beta_{A,B} = -4.2, -4.4, -4.6$, respectively and a fixed value of the scalar-tensor parameter $\alpha_{A,B} = 0.1$

The energy flux in the EOB coordinates is expressed as

$$\Phi_E^e = G_{\text{AB}} c^5 \Phi_E \ , \tag{6.50}$$

where Φ_E^e is the energy flux in the EOB coordinate system and Φ_E is the physical energy flux expressed in harmonic coordinates. As mentioned, using the coordinate transformation of Sect. 6.4 from harmonic coordinates to the EOB coordinates, we transform Φ_E to the EOB coordinates and find that the explicit expression for the energy flux in the EOB coordinates up to 1PN order reads

$$\Phi_E^e = \frac{4\nu^2}{15\tilde{r}_e^4}(\gamma_{AB}+2)\left(12\tilde{p}_e^2 - 11\tilde{p}_{r,e}^2\right) + \frac{1}{c^2}\left\{\frac{\nu^2}{105\tilde{r}_e^4}(\gamma_{AB}+2)\left[(449 - 852\nu + 336\gamma_{AB})\tilde{p}_e^4\right.\right.$$
$$+2(-1333 + 412\nu - 616\gamma_{AB})\tilde{p}_e^2\tilde{p}_{r,e}^2 + 3(687 - 4\nu + 280\gamma_{AB})\tilde{p}_{r,e}^4$$
$$-16\left(170 + 84\langle\bar{\beta}\rangle - 31\nu + 21\gamma_{AB}\right)\frac{\tilde{p}_e^2}{\tilde{r}_e} + 64\left(45 + 21\langle\bar{\beta}\rangle - 8\nu + 7\gamma_{AB}\right)\frac{\tilde{p}_{r,e}^2}{\tilde{r}_e}$$
$$\left.\left. + (16 - 64\nu)\frac{1}{\tilde{r}_e^2}\right]\right\}, \tag{6.51}$$

where $\tilde{r}_e = \frac{r_e}{G_{AB}M}$ is the dimensionless variable for the radial separation in EOB coordinates.

Similarly, the angular momentum flux in the EOB coordinates is expressed as

$$\Phi_J^e = c^5\frac{\Phi_J}{M}, \tag{6.52}$$

where Φ_J^e is the angular momentum flux in the EOB coordinate system and Φ_J is the physical angular momentum flux expressed in harmonic coordinates. Similarly, using the coordinate transformation of Sect. 6.4, we find that the explicit expression for the angular momentum in the EOB coordinates up to 1PN order reads

$$\Phi_J^e = \frac{4\nu^2}{5\tilde{r}_e^3}(\gamma_{AB}+2)\,\tilde{p}_{\phi,e}\left(2\tilde{p}_e^2 - 3\tilde{p}_{r,e}^2 + 2\frac{1}{\tilde{r}_e}\right) + \frac{1}{c^2}\left\{\frac{\nu^2\tilde{p}_{\phi,e}}{105\tilde{r}_e^3}(\gamma_{AB}+2)\left[(55 - 212\nu + 84\gamma_{AB})\tilde{p}_e^4\right.\right.$$
$$+6(-11 + 25\nu)\tilde{p}_e^2\tilde{p}_{r,e}^2 + 15(19 + 12\nu)\tilde{p}_{r,e}^4 - 4\left(247 + 84\langle\bar{\beta}\rangle + 137\nu + 42\gamma_{AB}\right)\frac{\tilde{p}_e^2}{\tilde{r}_e}$$
$$\left.\left.+4\left(480 + 168\langle\bar{\beta}\rangle + 25\nu + 147\gamma_{AB}\right)\frac{\tilde{p}_{r,e}^2}{\tilde{r}_e} - 2\left(661 - 86\nu + 336\langle\bar{\beta}\rangle + 126\gamma_{AB}\right)\frac{1}{\tilde{r}_e^2}\right]\right\}. \tag{6.53}$$

6.5.2 *Generic-Orbit Radiation Reaction Force*

The form of the general ansatz for the Schott terms is dictated by the fact that these terms enter due to the interaction with the radiation field, i.e., these are radiation effects. Therefore, we follow the postulates of constructing the radiation reaction acceleration given in Eq. (6.26) and additionally require that the angular momentum remains a pseudo-vector to construct the general ansatz. This implies that the Schott energy and angular momentum must be proportional to the odd powers in p_r, m/r

for the non-linear effects and ν^2 for vanishing radiation effects in the test mass limit. Hence, as an ansatz for the Schott terms we consider,

$$\mathrm{J}_{\mathrm{sct}}^{\mathrm{inst}} = \frac{\nu^2 \tilde{p}_{\phi,e}\tilde{p}_{r,e}}{\tilde{r}_e^2}\left[\alpha_1 + \frac{1}{c^2}\left(\alpha_2 \tilde{p}_{r,e}^2 + \alpha_3 \tilde{p}_e^2 + \frac{\alpha_4}{\tilde{r}_e}\right)\right] \tag{6.54}$$

$$E_{\mathrm{sct}}^{\mathrm{inst}} = \frac{\nu^2 \tilde{p}_{r,e}}{\tilde{r}_e^2}\left[\beta_1 \tilde{p}_{r,e}^2 + \beta_2 \tilde{p}_e^2 + \frac{\beta_3}{\tilde{r}_e} + \frac{1}{c^2}\left(\beta_4 \tilde{p}_{r,e}^4 + \beta_5 \tilde{p}_e^2 \tilde{p}_{r,e}^2 + \beta_6 \frac{\tilde{p}_{r,e}^2}{\tilde{r}_e} + \beta_7 \tilde{p}_e^4 + \beta_8 \frac{\tilde{p}_e^2}{\tilde{r}_e} + \frac{\beta_9}{\tilde{r}_e^2}\right)\right]. \tag{6.55}$$

Using the above ansatz of Schott terms along with the fluxes, we compute the radiation reaction force for the leading order (Newtonian contribution) and 1PN order. **Leading Order** At leading order, the radiation reaction force with the ansatz given in Eq. (6.54) and expanded in $\tilde{p}_{r,e}$ is

$$\mathscr{F}_r = \frac{\nu^2}{5\tilde{p}_{r,e}^2 \tilde{r}_e^3}\left(\tilde{p}_e^2 - \frac{1}{\tilde{r}_e}\right)\left[\tilde{p}_e^2\left(5\alpha_1 + 8\gamma_{\mathrm{AB}} - 5\beta_2 + 16\right) - 5\frac{\beta_3}{\tilde{r}_e}\right] + \mathscr{O}(\tilde{p}_{r,e}). \tag{6.56}$$

In the circular-orbit limit, $\tilde{p}_{r,e}$ is zero and therefore for a finite $\mathscr{F}_r$, we require that the $1/\tilde{p}_{r,e}$ term in the above equation is zero which leads to the solution

$$\beta_3 = 0, \qquad \alpha_1 = \frac{5\beta_2 - 16 - 8\gamma_{\mathrm{AB}}}{5}. \tag{6.57}$$

Now, expanding the circular limit condition $\mathscr{F}_r \tilde{p}_{\phi,e}/(\mathscr{F}_\phi \tilde{p}_{r,e}) - 1 = \mathscr{O}\left(\tilde{p}_{r,e}^2\right)$ in $\tilde{p}_{r,e}$ gives,

$$\frac{\mathscr{F}_r \tilde{p}_\phi}{\mathscr{F}_\phi \tilde{p}_r} - 1 = \frac{9\tilde{p}_e^2\left(5\beta_1 - 4\gamma_{\mathrm{AB}} - 8\right) - \left(44\gamma_{\mathrm{AB}} + 45\beta_1 + 30\beta_2 + 88\right)/\tilde{r}_e}{15\beta_2 \tilde{p}_e^2 + \left[48\left(\gamma_{\mathrm{AB}} + 2\right) - 15\beta_2\right]/\tilde{r}_e} + \mathscr{O}(\tilde{p}_{r,e}^2) \tag{6.58}$$

Again in the circular orbit limit, the first term in the above equation must vanish giving the solution

$$\beta_2 = -\frac{8}{3}\left(\gamma_{\mathrm{AB}} + 2\right), \qquad \beta_1 = \frac{4}{5}\left(\gamma_{\mathrm{AB}} + 2\right),$$
$$\alpha_1 = -\frac{64}{15}\left(\gamma_{\mathrm{AB}} + 2\right). \tag{6.59}$$

The leading order radiation reaction force in scalar-tensor theory then reads

$$\mathscr{F}_\phi = \frac{4\nu^2\left(\gamma_{\mathrm{AB}} + 2\right)\tilde{p}_{\phi,e}}{15\tilde{r}_e^3}\left(10\tilde{p}_e^2 - 39\tilde{p}_{r,e}^2 - \frac{22}{\tilde{r}_e}\right), \tag{6.60}$$

$$\mathscr{F}_r = -\frac{8\nu^2(\gamma_{\rm AB}+2)\,\tilde{p}_{r,e}}{15\tilde{r}_e^3}\left(\frac{11}{\tilde{r}_e} - 5\tilde{p}_e^2 + 12p_{r,e}^2\right) . \tag{6.61}$$

1 post-Newtonian Order Following the same steps taken for the leading order computation, we obtain that the radiation reaction force at 1PN order expanded in $p_{r,e}$ is

$$\begin{aligned}\mathscr{F}_r = {} & \frac{2\nu^2}{210\tilde{p}_{r,e}^2\tilde{r}_e^3}\left(\tilde{p}_e^2 - \frac{1}{\tilde{r}_e}\right)\Big[\frac{1}{\tilde{r}_e^2}\{-105\beta_9 - 16(-1+4\nu)(2+\gamma_{\rm AB}\} + 3\tilde{p}_e^4\{35\alpha_3 - 35\beta_7 \\ & +(2+\gamma_{\rm AB})(93+4\nu+28\gamma_{\rm AB}\} - \frac{\tilde{p}_e^2}{\tilde{r}_e}\{105\alpha_4 + 105\beta_8 + 2(2+\gamma_{\rm AB})(467+168\langle\bar{\beta}\rangle + 262\nu \\ & +210\gamma_{\rm AB})\}\Big] + \mathscr{O}(\tilde{p}_{r,e})\end{aligned} \tag{6.62}$$

Using the same argument used for the derivation of Eq. (6.57), we require that the $1/\tilde{p}_{r,e}$ term to be zero at 1PN order, and thus obtain the solution for the following unknown coefficients entering at 1PN order,

$$\begin{aligned}\beta_9 &= -\frac{16}{105}(4\nu-1)(\gamma_{\rm AB}+2) , \\ \beta_7 &= \frac{1}{35}\left(35\alpha_3 + 4\nu\gamma_{\rm AB} + 28\gamma_{\rm AB}^2 + 149\gamma_{\rm AB} + 8\nu + 186\right) , \\ \beta_8 &= \frac{1}{105}\left(105\alpha_4 - 336\beta\gamma_{\rm AB} - 524\nu\gamma_{\rm AB} - 420\gamma_{\rm AB}^2 - 1774\gamma_{\rm AB} - 672\langle\bar{\beta}\rangle - 1048\nu - 1868\right) .\end{aligned} \tag{6.63}$$

The expansion of the circular limit condition $\mathscr{F}_r\tilde{p}_{\phi,e}/(\mathscr{F}_\phi\tilde{p}_{r,e}) - 1 == \mathscr{O}\left(\tilde{p}_{r,e}^2\right)$ in $\tilde{p}_{r,e}$ at 1PN order gives,

$$\frac{\mathscr{F}_r\tilde{p}_{\phi,e}}{\mathscr{F}_\phi\tilde{p}_{r,e}} - 1 = -\frac{\mathscr{N}(\tilde{p}_e,\tilde{r}_e)}{\mathscr{D}(\tilde{p}_e,\tilde{r}_e)} + \mathscr{O}(\tilde{p}_{r,e}^2) , \tag{6.64}$$

where

$$\begin{aligned}\mathscr{N}(\tilde{p}_e,\tilde{r}_e) = {} & 9\tilde{p}_e^4(-35\alpha_2 + (\gamma_{AB}+2)(-28\gamma_{AB}+54\nu-11) + 35\beta_5) + \frac{\tilde{p}_e^2}{\tilde{r}_e}(-210\alpha_3 \\ & +315(\alpha_2 - \beta_5 + \beta_6) + 2(2+\gamma_{AB})(-822+336\langle\bar{\beta}\rangle + 637\nu - 826\gamma_{AB})) \\ & + \frac{1}{\tilde{r}_e^2}\left(-210\alpha_4 - 315\beta_6 + 4(2+\gamma_{AB})(1552+350\langle\bar{\beta}\rangle + 291\nu + 784\gamma_{AB})\right) ,\end{aligned} \tag{6.65}$$

$$\begin{aligned}\mathscr{D}(\tilde{p}_e,\tilde{r}_e) = {} & -3\tilde{p}_e^4(35\alpha_3 + (2+\gamma_{AB})(28\gamma_{AB}+4\nu+93)) \\ & + \frac{\tilde{p}_e^2}{\tilde{r}_e}\left(105\alpha_3 - 105\alpha_4 + 4(2+\gamma_{AB})\left(42\gamma_{AB} + 84\langle\bar{\beta}\rangle + 305\nu + 79\right)\right) \\ & + \frac{1}{\tilde{r}_e^2}\left(105\alpha_4 + 2(2+\gamma_{AB})\left(-322\gamma_{AB} + 784\langle\bar{\beta}\rangle - 310\nu + 437\right)\right) .\end{aligned} \tag{6.66}$$

In the circular-orbit limit, the first term in the above expansion vanishes giving the solution,

$$\alpha_2 = \frac{1}{35}\left(54\nu\gamma_{\rm AB} - 28\gamma_{\rm AB}^2 - 67\gamma_{\rm AB} + 35\beta_5 + 108\nu - 22\right)$$

$$
\begin{aligned}
\alpha_4 &= \frac{1}{210}\left(-210\alpha_3 + (2+\gamma_{AB})(4465 + 2072\langle\bar{\beta}\rangle + 2924\nu + 1232\gamma_{AB})\right) \\
\beta_6 &= \frac{2\alpha_3}{3} - \frac{(\gamma_{\rm AB}+2)}{315}\left(-1904\gamma_{\rm AB} + 672\langle\bar{\beta}\rangle + 1760\nu - 1743\right) \\
\beta_8 &= \frac{1}{30}(\gamma_{\rm AB}+2)\left(56\gamma_{\rm AB} + 200\langle\bar{\beta}\rangle + +268\nu + 371\right) - \alpha_3
\end{aligned}
\tag{6.67}
$$

From Eqs. (6.63) and (6.67), we observe that the solution of the unknown coefficients entering the ansatz at 1PN order is given in terms of 3 arbitrary coefficients, i.e. out of 9 unknown coefficients 6 coefficients are determined in terms of 3 arbitrary coefficients. Since the 3 coefficients are arbitrary, for simplification we set these coefficients to zero, namely $\alpha_3 = \beta_5 = \beta_4 = 0$. Using this result, the 1PN order contribution to the radiation reaction force becomes,

$$
\begin{aligned}
\mathscr{F}_\phi &= \frac{\nu^2\tilde{p}_{\phi,e}(2+\gamma_{AB})}{210c^2\tilde{r}_e^3}\Big[60\,(21\nu - 15 - 14\gamma_{AB})\,\tilde{p}_{r,e}^4 - 6\,(93 + 4\nu + 28\gamma_{AB})\,\tilde{p}_e^4 \\
&+2\left(9599 + 3472\langle\bar{\beta}\rangle + 4890\nu + 2968\gamma_{AB}\right)\frac{\tilde{p}_{r,e}^2}{\tilde{r}_e} - \left(3833 + 1400\langle\bar{\beta}\rangle + 484\nu + 896\gamma_{AB}\right)\frac{\tilde{p}_e^2}{\tilde{r}_e} \\
&+18\,(93 + 4\nu + 28\gamma_{AB})\,\tilde{p}_{r,e}^2\tilde{p}_e^2 + \left(6213 + 1684\nu + 5208\langle\bar{\beta}\rangle - 56\gamma_{AB}\right)\frac{1}{\tilde{r}_e^2}\Big], \\
\mathscr{F}_r &= \frac{\nu^2\tilde{p}_{r,e}(2+\gamma_{AB})}{210c^2\tilde{r}_e^3}\Big[60\,(21\nu - 15 - 14\gamma_{AB})\,\tilde{p}_{r,e}^4 - 6\,(93 + 4\nu + 28\gamma_{AB})\,\tilde{p}_e^4 \\
&+4\left(3958 + 1064\langle\bar{\beta}\rangle + 691\nu + 1414\gamma_{AB}\right)\frac{p_{r,e}^2}{\tilde{r}_e} - \left(3833 + 1400\langle\bar{\beta}\rangle + 484\nu + 896\gamma_{AB}\right)\frac{\tilde{p}_e^2}{\tilde{r}_e} \\
&-6\,(198\nu - 429 - 224\gamma_{AB})\,\tilde{p}_{r,e}^2\tilde{p}_e^2 + \left(6213 + 1684\nu + 5208\langle\bar{\beta}\rangle - 56\gamma_{AB}\right)\frac{1}{\tilde{r}_e^2}\Big].
\end{aligned}
\tag{6.68}
$$

6.6 Discussion

The waveform models used to construct template banks for GW detection have been extended from quasi-circular to eccentric orbits in GR [13, 14] as with the new detectors the eccentricity might be observed soon and must be included in the parameter estimation of GW signals. In this chapter, we have extended the quasi-circular orbit limit analytical results for the radiation reaction force for scalar-tensor theories to generic orbits for nonspinning binaries without tides,i.e. the eccentric orbits.

In particular, we have derived the coordinate transformations required to map the energy and angular momentum flux derived in harmonic coordinates to EOB coordinates up to 1PN order and then derived the radiation reaction force for generic orbits up to 1PN order by only considering the tensorial contribution to the fluxes. Since the radiation reaction effects start at 1.5PN order for scalar-tensor theories, we have also discussed the coordinate transformation from harmonic coordinates to

ADM-like coordinates by investigating the alternative gauge choices for the radiation reaction term following Ref. [15, 16]. Unlike in GR [15, 16], we find that there is no gauge freedom in the radiation reaction effects at the leading order for scalar-tensor theories.

The next steps of our work will include recasting the results obtained here in the form of the quasi-circular radiation reaction force employed in TEOBResumS or SEOBNR for scalar-tensor theories and calculating the waveform modes for eccentric binaries. Following this, we will incorporate the radiation reaction force and the waveform modes in the inspiral-merger-ringdown quasi-circular orbit TEOBResumS waveform model for scalar-tensor theories.

References

1. Jain T, Rettegno P (2024) Radiation reaction force for scalar-tensor theories in effective-one-body formalism, in preparation
2. Julié F-L (2018) Reducing the two-body problem in scalar-tensor theories to the motion of a test particle : a scalar-tensor effective-one-body approach. Phys Rev D 97:024047. https://doi.org/10.1103/PhysRevD.97.024047, arXiv:1709.09742 [gr-qc]
3. Julié FL, Deruelle N (2017) Two-body problem in scalar-tensor theories as a deformation of general relativity : an effective-one-body approach. Phys Rev D 95:124054. https://doi.org/10.1103/PhysRevD.95.124054, arXiv:1703.05360 [gr-qc]
4. Julié F-L, Baibhav V, Berti E, Buonanno A (2023) Third post-Newtonian effective-one-body Hamiltonian in scalar-tensor and Einstein-scalar-Gauss-Bonnet gravity. Phys. Rev. D 107:104044 (2023). https://doi.org/10.1103/PhysRevD.107.104044, arXiv:2212.13802 [gr-qc]
5. Sennett N, Marsat S, Buonanno A (2016) Gravitational waveforms in scalar-tensor gravity at 2PN relative order. Phys Rev D 94:084003. https://doi.org/10.1103/PhysRevD.94.084003, arXiv:1607.01420 [gr-qc]
6. Sperhake U, Berti E, Cardoso V, González JA, Brügmann B, Ansorg M (2008) Eccentric binary black-hole mergers: The Transition from inspiral to plunge in general relativity. Phys Rev D 78:064069. https://doi.org/10.1103/PhysRevD.78.064069, arXiv:0710.3823 [gr-qc]
7. Hinder I, Vaishnav B, Herrmann F, Shoemaker DM, Laguna P (2008) Universality and final spin in eccentric binary black hole inspirals. Phys Rev D 77:081502. https://doi.org/10.1103/PhysRevD.77.081502, arXiv:0710.5167 [gr-qc]
8. Sperhake U, Cardoso V, Pretorius F, Berti E, Gonzalez JA (2008) The High-energy collision of two black holes. Phys. Rev. Lett. 101:161101 arXiv:0806.1738 [gr-qc]
9. Shibata M, Okawa H, Yamamoto T (2008) High-velocity collision of two black holes. Phys Rev D 78:101501. https://doi.org/10.1103/PhysRevD.78.101501. arXiv:0810.4735 [gr-qc]
10. Bernard L, Blanchet L, Trestini D (2022) Gravitational waves in scalar-tensor theory to one-and-a-half post-Newtonian order. JCAP 08(08):008. https://doi.org/10.1088/1475-7516/2022/08/008. arXiv:2201.10924 [gr-qc]
11. Damour T, Esposito-Farese G (1992) Tensor multiscalar theories of gravitation. Class Quant Grav 9:2093–2176 https://doi.org/10.1088/0264-9381/9/9/015
12. Damour T, Esposito-Farese G (1996) Testing gravity to second postNewtonian order: A Field theory approach. Phys. Rev. D 53:5541–5578. https://doi.org/10.1103/PhysRevD.53.5541
13. Bini D, Damour T (2012) Gravitational radiation reaction along general orbits in the effective one-body formalism. Phys Rev D 86:124012 arXiv:1210.2834 [gr-qc]
14. Khalil M, Buonanno A, Steinhoff J, Vines J (2021) title Radiation-reaction force and multipolar waveforms for eccentric, spin-aligned binaries in the effective-one-body formalism. Phys Rev D 104:024046. https://doi.org/10.1103/PhysRevD.104.024046. arXiv:2104.11705

15. Iyer BR, Will CM (1993) Post Newtonian gravitational radiation reaction for two-body systems. Phys Rev Lett 70:113–116. https://doi.org/10.1103/PhysRevLett.70.113
16. Iyer BR, Will CM (1995) PostNewtonian gravitational radiation reaction for two-body systems: Nonspinning bodies. Phys Rev D 52:6882–6893. https://doi.org/10.1103/PhysRevD.52.6882
17. Gopakumar A, Iyer BR, Iyer S (1997) Second post Newtonian gravitational radiation reaction for two-body systems: nonspinning bodies. Phys Rev D 55:6030–6053. https://doi.org/10.1103/PhysRevD.57.6562. [Erratum: Phys.Rev.D 57, 6562 (1998)], arXiv:gr-qc/9703075
18. Schott G (1915) Vi. on the motion of the lorentz electron, The London, Edinburgh, and Dublin Philos Mag J Sci 29:49–62. https://doi.org/10.1080/14786440108635280
19. Jain T, Rettegno P (2024) Angular momentum flux in scalar-tensor theories up to 1.5 post-newtonian order. arXiv:2407.10908v1
20. Damour T, Schäfer G (1985) Lagrangians for n point masses at the second post-Newtonian approximation of general relativity. Gen Rel Grav 17:879–905. https://doi.org/10.1007/BF00773685
21. Jain T, Rettegno P, Agathos M, Nagar A, Turco L (2023) Effective-one-body Hamiltonian in scalar-tensor gravity at third post-Newtonian order. Phys Rev D 107:084017. https://doi.org/10.1103/PhysRevD.107.084017, arXiv:2211.15580 [gr-qc]
22. Lange J et al (2017) Parameter estimation method that directly compares gravitational wave observations to numerical relativity. Phys Rev D 96:104041. https://doi.org/10.1103/PhysRevD.96.104041, arXiv:1705.09833 [gr-qc]
23. Lange J, O'Shaughnessy R, Rizzo M (2018) Rapid and accurate parameter inference for coalescing, processing compact binaries. arXiv:1805.10457 [gr-qc]

Part II
Inference of Source Properties: Neutron Stars

Chapter 7
Improving Inference on Neutron Star Properties Using Information from Binary Merger Remnants

We perform an analysis of the binary neutron star coalescence event GW170817, to improve the inference of the properties of neutron stars. The observation of electromagnetic counterparts of the GW170817 event suggests that the event did not lead to the prompt formation to a black-hole. In this work, we first briefly overview different numerical techniques used to perform the numerical relativity simulations of binary-neutron stars followed by a review of the mass-ratio dependent fits for the threshold mass ($M_{\rm thr}$), a characteristic mass used to classify the merger remnant. Finally, we classify the GW170817 LIGO-Virgo data sample into the prompt collapse to a black-hole using the q-dependent threshold mass fits to infer the properties of neutron stars.

This is based on the work presented in Ref. [1].

7.1 Introduction

In 2015, the first gravitational-wave (GW) detection from the merger of a binary black-hole (BBH) [2] by the LIGO-Virgo detectors [3, 4] opened a new era of GW astronomy. Subsequently, LIGO-Virgo detected GW signals from the coalescence of binary neutron-star (BNS) systems, namely GW170817 [5–7] and GW190425 [8].

The merger of a BNS system can lead either to a prompt collapse to a BH or a formation of a NS remnant (non-prompt collapse) [9–15], that can be either stable or unstable, depending on its mass and spin. The detection of electromagnetic (EM) counterparts from space-based or ground-based observatories is crucial to identify the signal from the BNS merger as a non-prompt collapse signal. The EM radiation can be emitted either from the dynamical ejection of neutron-star matter or from the formation of a disk surrounding the remnant. The follow-up searches after the detection of the GW signal led to the observation of EM counterparts in the case of GW170817, while in the case of GW190425, no EM signal was detected. The EM

T. Jain, *Surfing the Gravitational Waves*, Springer Theses,
https://doi.org/10.1007/978-3-031-96180-9_7

counterpart of GW170817, known as AT2017gfo, consisted of a short gamma-ray burst (sGRB) [16–18] followed up by strong emission from X-ray to radio [19–24], that was compatible with a kilonova transient event, and pointed to a non-prompt collapse merger. The lack of detection of the EM counterparts for GW190425 could be due to its poor sky-localization (8284 $\deg^2$ as its 90% confidence region, compared to 28 $\deg^2$ for GW170817), and its large distance (159^{+69}_{-71}Mpc compared to $\sim$ 40Mpc for GW170817), as it was very unlikely that a low-latency survey would have picked up such a faint source while having to cover a wide patch. It is also possible that the EM emission was highly collimated and that our line of sight lied outside the source's emission cone.

Over the years, various numerical relativity (NR) studies [11–13, 25–28] investigated the remnants of BNS mergers, concluding that the coalescence of a BNS system is followed by the prompt collapse to a BH if the total mass of the binary is greater than some characteristic mass called the *threshold mass*. Thus, accurate modelling of this threshold mass is important as it characterizes the outcome of the merger remnant for a BNS of a given total mass. The first use of the non-prompt collapse hypothesis for GW170817 was made in Ref. [29] to place a bound on the NS radius, $R_{1.6} > 10\,\mathrm{km}$, for a fiducial NS mass of 1.6 $M_\odot$.

Recent works, Ref. [30], hereafter Bauswein et al., and Ref. [31], hereafter Kölsch et al., provide empirical formulae for the threshold mass, as a function of the equation of state (EOS) of NS matter and the mass ratio of the BNS, based on NR simulations. In this work, we make use of these fits and the presence of an EM counterpart for the GW170817 event to improve the inference on the properties of NS matter. On applying this method to GW170817 we find that a significant part of the parameter space is not supported by the observation of the EM counterpart.

The chapter is organised as follows. In Sect. 7.2, we briefly discuss NR simulation techniques used in Refs. [30–32] and the range of EOSs and mass-ratios q used in these studies. Section 7.3 summarises the numerical fits of threshold mass, the impact of the mass-ratio on threshold mass, the method based on which we classify the outcome of the BNS merger, and the Bayesian analysis used for GW170817. Finally, we present our results for the GW170817 signal, along with the results from the injection studies of Ref. [32], hereafter Agathos et al., in Sect. 7.4 with concluding remarks in Sect. 7.5.

7.2 Overview of Numerical Simulations of Neutron Star Binaries

In this section, we discuss the numerical techniques and the set-up of the BNS systems used in Bauswein et al. [30] and Kölsch et al. [31] to study the impact of mass-ratio on the onset of a prompt collapse to a BH.

The different groups use independent NR codes to simulate binaries for their analysis. The NR simulations of Kölsch et al. use the `BAM` code for dynamical evolutions which is based on the 3+1 decomposition of the Einstein's equation in the Z4c formalism [33, 34]. The $(1 + \log)$ slicing and gamma-driver gauge conditions are used for lapse and shift, respectively [35–37]. BAM uses nested Cartesian grids to solve both the general relativistic hydrodynamics (GRHD) equations as well as Einstein's equations. On the other hand, the smoothed particle hydrodynamics (SPH) code [38–40] used by Bauswein et al. employs the conformally flat approximation of general relativity to solve Einstein's equations [41, 42]. The SPH code uses the Lagrangian particle method to solve the equations of GRHD [43] and the multigrid solver to solve the field equations. The particle mesh codes are then used to translate between the SPH particles and the gravity grid [44]. BAM utilises AMR (adaptive mesh refinement) techniques [45] whereas the multilevel adaptive technique [46] is used in SPH for solving nested grid setup.

For their analysis, Kölsch et al. considers 290 NR simulations with varying total mass, and mass-ratios $0.57 \leqslant q \leqslant 1$. They consider a restricted set of 3 EOSs, ALF2 [47], SLy [48], and H4 [49] such that their maximum mass is compatible with the observation of binary pulsar PSR J0348+0432, i.e. $M_{\rm max}^{\rm TOV} \geq 1.97\,{\rm M}_\odot$ where $M_{\rm max}^{\rm TOV}$ is the maximum mass of the non-rotating NSs [50]. The maximum mass of non-rotating NSs corresponding to these EOSs are $1.99\,{\rm M}_\odot$, $2.06\,{\rm M}_\odot$, and $2.03\,{\rm M}_\odot$, respectively. However, as mentioned in Kölsch et al. due to a small set of EOSs, the $M_{\rm thr}$ fit based only on their data does not reliably predict $M_{\rm thr}$ for other NR simulated EOSs given in Perego et al. [51]. Therefore, as in Kölsch et al. we will consider a fit derived from the data of Kölsch et al. as well as Perego et al. [31, 51, 52] (see, Table X of Kölsch et al. [31]) for our analysis. On the other hand, Bauswein et al. considers a total of 40 EOS models, out of which 23 are hadronic EOS models which they refer to as their "base sample" [47, 48, 53–71], 8 belong to the "excluded hadronic sample" of EOSs [49, 63–66, 72–74] and the remaining 9 are EOSs which include a phase transition to deconfined quark matter called "hybrid" EOS models [59, 75–80]. They consider three mass-ratios, namely 1.0, 0.85 and 0.7. Based on the classification of the EOS models, Bauswein et al. presents the threshold mass fit formula for each EOS model as well as for the combinations of EOS models. As the EOSs considered in Kölsch et al. and Perego et al. are included in the base and excluded EOS models of Bauswein et al., we only consider the threshold mass fit formula for the "base+excluded" model for our analysis of the GW170817 signal.

To verify the robustness of the two q-dependent threshold mass fits presented in Bauswein et al. and Kölsch et al., we use these fits on the 17 NR simulations considered in Agathos et al. [32]. In that study, the authors used previously presented NR simulation data from Refs. [81–87] along with two NR simulations performed with the WhiskyTHC code [88–90]. The study considers mass-ratios $0.86 \leqslant q \leqslant 1.0$ with a total of 8 EOSs, out of which 5 are microphysical EOSs: BHB$\Lambda\phi$ [53], DD2 [56, 57], LS220 [63], SFHo [67], SLy-SOR [91], and the remaining 3 are piecewise polytropic EOSs: ALF2, 2B, and SLy [71].

7.3 Methods

7.3.1 Numerical Fits for Remnant Classification

The merger of a BNS system can lead either to the prompt collapse to a BH or the formation of a NS remnant. In NR simulations, it has been observed that a high total-mass binary leads to a prompt collapse to a BH whereas a low total-mass binary with similar BNS configuration leads to a meta-stable remnant which may undergo a delayed collapse to a BH. Therefore, a characteristic mass M_{thr} can be defined to reliably classify the merger remnant in the NS binaries, such that, if the total mass (M) of the NS binary is larger than the M_{thr} then the merger product is a prompt collapse to a BH. The M_{thr} depends on the EOS, mass-ratio, and spins as explained below.

The dynamics of the BNS system is dependent on the nuclear matter, i.e. the unknown EOS and the binary masses, therefore, the merger product is also determined by the EOS for the particular BNS configuration. The dependency of the merger product and hence, the threshold mass on the EOS can then be expressed by the EOS-dependent parameters like maximum-mass, maximum-radius, tidal deformability, etc., of the non-rotating NS, which are uniquely linked to the EOS. The fitting formula for M_{thr} is derived using the large sets of NR hydrodynamical simulations [12, 13] and is given by,

$$M_{\mathrm{thr}} = k_{\mathrm{thr}} M_{\mathrm{max}}^{\mathrm{TOV}} \,. \tag{7.1}$$

Here, k_{thr} depends on the EOS, the mass-ratio and the spins, and $M_{\mathrm{max}}^{\mathrm{TOV}}$ is the mass of the heaviest stable non-rotating NS supported by the EOS. The effect of mass-ratio and spin on the formation of the merger remnant and hence, M_{thr} is discussed below using Newtonian point particle dynamics.

For a sample of 12 hadronic EOSs and equal-mass binaries, Ref. [13] observed an approximate EOS-independent linear relation between k_{thr} and the maximum compactness (C_{max}) of the non-rotating NS. For a fixed mass-ratio of $q = 0.7$, the study of mass-ratio effects on the threshold mass indicates a monotonic relation between the two parameters [28], i.e., the threshold mass decreases with the increase in the asymmetry. A fitting formula for M_{thr} dependent on the EOS and the mass-ratio was first derived by Bauswein et al. [30], shortly afterwards followed by an independent study by Kölsch et al. [31], who also pointed out that not all EOSs give a monotonically increasing threshold mass with increasing mass ratio.

This effect of the mass-ratio on the threshold mass can be explained using Newtonian point particle dynamics. The stability of the remnant formed in a binary merger system is determined by two forces acting in opposite directions: i) the destabilization effect of the mass, i.e., the attractive force due to gravity towards the centre, and (ii) the centrifugal support, i.e., the centrifugal force acting away from the centre. Now, from Kepler's law for a binary system with no eccentricity and $m_1 > m_2$, the angular momentum of each component $\big(L(q)_{1,2}\big)$ and the total angular momentum $L(q) = L_1(q) + L_2(q)$ in terms of $q = \frac{m_2}{m_1}$ are given by,

$$L_1(q) = \frac{q^2}{(1+q)^3} M\sqrt{aM}\,, \qquad L_2(q) = \frac{q}{(1+q)^3} M\sqrt{aM}\,, \qquad L(q) = \frac{q}{(1+q)^2} M\sqrt{aM}\,, \tag{7.2}$$

where a is the orbital separation and $M = m_1 + m_2$ is the total mass. From the above formulae for the angular momentum, we observe that the angular momentum of the less massive component (secondary) of the binary is greater than the angular momentum of the more massive component (primary),i.e. $L_2(q) > L_1(q)$ for all the values of mass-ratio $0 < q < 1$.

In an asymmetric binary system, the secondary component distorts and wraps around the primary binary component, therefore, the angular momentum at the centre of the remnant is less in asymmetric binaries compared to the angular momentum of equal-mass binaries. Furthermore, from the above equation for the total angular momentum $L(q)$, we observe that the angular momentum decreases with a decrease in q, i.e., decreases with the increase in asymmetry for all values of q ranging from $0 < q < 1$. The decrease in both the total angular momentum and the angular momentum at the centre of the remnant suggests a decrease in the centrifugal force with an increase in the asymmetry for fixed total mass. This implies that for two binaries with the same orbital separation at the point of merging (the point at which the angular momentum of the remnant is calculated) and the same total mass, the asymmetric binary would have less centrifugal support compared to the symmetric binary. Therefore, the stability due to the centrifugal support decreases for asymmetric binaries suggesting a decrease in the threshold mass with the increase in asymmetry.

The effect of spin can also be discussed using a similar argument. Consider a BNS system with the spin of the primary component aligned (anti-aligned) with the angular momentum and a non-rotating secondary component. For this system, the centrifugal support increases (decreases) suggesting a larger threshold mass with the increase (decrease) in the spin for the aligned (anti-aligned) spin cases[1]. However, in our current work, we do not consider the effect of spins and use the fitting formula derived in Refs. [30, 31] dependent only on EOS and mass-ratio.

The non-monotonicity observed in Refs. [30, 31] between the threshold mass and the mass-ratio can be explained by the observation of Ref. [28] that the mass-ratio effect on the threshold mass is also dependent on the EOS configuration. This effect could also be explained using a similar argument as above if we assume that the orbital separation at the merger is calculated using $a = r_1(m_1) + r_2(m_2)$, i.e., is dependent on the mass-radius relation unique to the EOS. The dependency of a on the EOS consequently implies that the angular momentum and hence the mass-ratio effect on the threshold mass is EOS dependent.

Now, for our analysis, we consider the $M_{\rm thr}\left(M_{\rm max}^{\rm TOV}, R_{\rm max}, q\right)$ fit from Bauswein et al. ("Bauswein+" from here on), which we denote by $M_{\rm thr}^{B}$ and explicitly expressed in Eq. (7.3), and the $M_{\rm thr}\left(M_{\rm max}^{\rm TOV}, R_{1.6\,{\rm M}_\odot}, q\right)$ fit from Kölsch et al. ("Kölsch+" from here on), which we denote by $M_{\rm thr}^{K}$ and explicitly expressed in Eq. (7.4). In the above equations, $M_{\rm max}^{\rm TOV}$, $R_{\rm max}$, $R_{1.6\,{\rm M}_\odot}$ are the maximum mass, radius at the maximum mass

[1] In Ref. [92], Tootle et al. were the first to derive a fitting formula for the threshold mass that also includes dependence on the NS spins; we shall not consider the effect of spins in the current work.

and radius at 1.6 $M_\odot$ of the non-rotating NS, respectively. We use the Bauswein+ and Kölsch+ numerical fits of $M_{\rm thr}$, as they better capture their NR results than the other fits given there. However, we note an exception of a more fine-tuned fit by Kölsch et al. which instead of some fiducial radius, depends on the estimated value of threshold mass at $q = 1$ $\left(M_{\rm thr}^{q=1}\right)$ based on the NR simulations. However, as $M_{\rm thr}^{q=1}$ is a quantity that can only be estimated by NR simulations for a given EOS, we cannot use this fit for our GW170817 analysis, where the EOS is being sampled over; we can only use it in our NR injection studies of Agathos et al. [32] (see Sect. 7.4.2). We denote this fit by $M_{\rm thr}^{K1}$ and show its explicit expression in Eq. (7.5). In summary, we thus have,

$$\begin{aligned} M_{\rm thr}^{B}(M_{\rm max}^{\rm TOV}, R_{\rm max}, q) =& -0.07794 - 5.011(1-q)^3 + \left[0.1863 - 1.970(1-q)^3\right] R_{\rm max} \\ &+ \left[0.497 + 11.098(1-q)^3\right] M_{\rm max}^{\rm TOV}, \qquad (7.3) \\ M_{\rm thr}^{K}(M_{\rm max}^{\rm TOV}, R_{1.6M_\odot}, q) =& \, 0.246 + \left[0.463 + 0.735(1-q) + 0.172(1-q)^3\right] M_{\rm max}^{\rm TOV} \\ &+ \left[0.141 - 0.116(1-q) - 0.214(1-q)^3\right] R_{1.6M_\odot}, \qquad (7.4) \\ M_{\rm thr}^{K1}(M_{\rm thr}^{q=1}, M_{\rm max}^{\rm TOV}, q) =& \, 0.281 - \left[0.03667 - 1.11(1-q) - 1.191(1-q)^3\right] M_{\rm max}^{\rm TOV} \\ &+ \left[0.932 - 0.753(1-q) - 1.625(1-q)^3\right] M_{\rm thr}^{q=1}. \qquad (7.5) \end{aligned}$$

We will also compare the predictions of the q-dependent $M_{\rm thr}$ fits listed above with the q independent $M_{\rm thr}$ fits used in Agathos et al.. The latter were derived as a relation between $k_{\rm thr}$ and $C_{\rm max}$,

$$k_{\rm thr}(C_{\rm max}) = -(3.29 \pm 0.23)C_{\rm max} + (2.392 \pm 0.064), \qquad (7.6)$$

by combining the results of Refs. [13, 93] with data by the CoRe collaboration [82, 94].

7.3.2 *Bayesian Analysis of Neutron Star Binaries*

We perform Bayesian analysis on the GW detector data using a waveform model with matter effects that depend on the masses and the EOSs, to recover posterior samples of compact binary coalescence (CBC) parameters of the signal GW170817. In addition to the standard CBC parameters for BBHs—masses, spins, distance, coalescence time, phase, sky location and orientation angles—we also treat the EOS as an unknown function that we need to sample over. The EOSs are defined by the relation between pressure and density $p(\rho)$ which can be empirically modelled as spectral expansion (see, Ref. [95] for review). We use the four-dimensional family of EOSs spanned by the pressure based spectral parametrization [96–98] where the adiabatic index $\Gamma(p)$, which characterises the stiffness of the equation of state at given densities, is approximated by a third-order polynomial $\Gamma(p) = \exp\left(\Sigma_{k=0}^{3} \gamma_k \log(p/p_0)\right)$.

The γ_in's are the spectral parameters which determine the particular adiabatic index $\Gamma(p)$. The adiabatic index can then be inverted to compute the energy and the rest-mass density, and thus, the particular EOS [96]. Therefore, we sample over the additional parameters (γ_0, γ_1, γ_2, γ_3), to construct the EOS for the NS matter and derive all EOS-dependent parameters, such as tidal deformability, NS radii, etc. by solving the TOV equations for the sampled values of m_1 and m_2.

An obvious benefit of using this method is that for each point in the parameter space of (γ_0,γ_1, γ_2, γ_3), we can also calculate global properties of the EOS parameters like $M_{\rm max}$, $R_{\rm max}$, and $R_{1.6{\rm M}_\odot}$. We can then use these in our fits (Eqs. (7.3) and (7.4)) to compute the threshold mass $M_{\rm thr}$ for each sample and thus translate the joint posterior of (m_1, m_2, γ_0, γ_1, γ_2, γ_3) to the joint posterior probability of ($M_{\rm thr}$, M), where M is the total mass of the BNS system.

Finally, we classify each sample as either a prompt collapse to a BH or a NS remnant (non-prompt collapse), based on whether M is greater or less than the threshold mass. As the observation of EM counterparts after the observation of the GW170817 signal indicates that it was most likely a non-prompt collapse signal, we can reject a posteriori that part of the BNS parameter space for which our calculation predicts a prompt collapse. We can then compare the resulting posteriors with the original results from the LIGO-Virgo Collaboration (LVC) analyses [6, 98], and assess the effect of our truncating procedure on measuring the properties of the BNS source.

We also repeat the analysis described above by imposing a hard lower bound on the maximum non-rotating TOV mass of the NS EOS, to be consistent with the observation of the binary pulsar PSR J0348+0432[2] [50]. The measured mass of the pulsar is $M_{J0348+0432} = 2.01 \pm 0.04$ M$_\odot$, so we impose a conservative constraint on the maximum TOV mass as $M_{\rm max}^{\rm TOV} \geq 1.97$ M$_\odot$. This effectively removes a non-negligible part of the EOS parameter space, in particular the subset of models that are too soft.[3] to support NSs at masses larger than 1.97 M$_\odot$.[4]

In Sect. 7.4, we will compare the improvement in parameter estimation with and without this constraint.

---

[2] Here we use the measured mass $M_{\rm NS} = 2.01^{+0.4}_{-0.4}$ M$_\odot$ of the binary pulsar PSR J0348+0432 to impose a lower bound on the maximum mass instead of the measured mass $M_{\rm NS} = 2.35^{+0.17}_{-0.17}$ M$_\odot$ of the most massive known binary pulsar PSR J0952-0607 [99]. This is because in our analysis, we are imposing a lower bound on the maximum mass of a non-rotating neutron star and out of all the known massive pulsars, i.e. with $M_{\rm NS} \geq 2.0$ M$_\odot$, PSR J0348+0432 is the pulsar with the slowest rotation (spin frequency$\sim$ 25 Hz) [99, 100]. Therefore, as a conservative estimate we use the measured mass of PSR J0348+0432 to impose a lower bound on the maximum mass of a non-rotating neutron star.

[3] The soft EOSs are the EOSs that support small increase in the pressure with change in the density which in terms of mass-radius relation implies a small radius for the neutron star.

[4] From a similar argument, we would expect that if we use the measured mass $M_{\rm NS} = 2.35^{+0.17}_{-0.17}$ M$_\odot$ of the most massive known binary pulsar PSR J0952-0607 as a lower bound on the maximum mass for GW170817 data analysis, it would further remove the soft part of the EOS parameter space. Therefore, it will be an interesting analysis for the future to quantitatively study the effect of using this mass as a lower bound on the maximum mass. The analysis could be done by first generating posteriors using this constraint in the prior, which is beyond the scope of our current work.

The `IMRPhenomPv2_NRTidalv1` waveform model was used to measure the source parameters of GW170817 for both the parameter estimation studies of Ref. [6] and the EOS studies of Ref. [98]. The BBH baseline in `IMRPhenomPv2_NRTidalv1` model rests on the point-particle model presented in Ref. [101], which incorporates effective-one-body (EOB) and NR-tuned.[5] tidal effects [102, 103], spin-induced quadrupole effects [105–109] and precession effects. Regarding the matter sector, Ref. [98] makes minimal assumptions about the nature of the source and samples over the tidal deformability parameters in contrast to Ref. [6] where the condition that both NSs in the binary are described by the same EOS is imposed and the 4-dimensional spectral decomposition is used to sample the space of EOSs. The two studies of Refs. [6, 98] also compare the results between different waveform models, namely `SEOBNRv4_ROM_NRTidal`[110], `IMRPhenomD_NRTidal` [111], and `TaylorF2` and found that the systematic uncertainties due to different waveform models are lower than the statistical uncertainties in observing the GW170817 signal.[6]

7.3.3 Validation of Threshold Mass Fits

In addition to the GW170817 re-analysis, we implement the above Bayesian analysis on the injection study of the BNS systems given in Agathos et al.. From the obtained joint posterior probability of (M_{thr}, M), we classify the merger outcome into either prompt collapse to BH or non-prompt collapse, and compute the posterior probability of prompt collapse, given by

$$P_{\mathrm{PC}} = P(M > M_{\mathrm{thr}} | d, \mathcal{H}_{\mathrm{PC}}), \tag{7.7}$$

for each BNS system of Agathos et al.. Here, by d we denote the simulated data of the detector network (in this case from LIGO Livingston (L1), LIGO Hanford (H1) and Virgo (V1) at design sensitivity), and by $\mathcal{H}_{\mathrm{PC}}$ we denote the underlying model that we use for estimating the prompt collapse threshold for the total mass. We then also run our above analysis with the additional maximum mass constraint from PSR J0348+0432 and observe the effect this has on the estimate of P_{PC}.

[5] It should be noted that the tidal effects in `IMRPhenomPv2_NRTidalv1` are tuned with the NR simulations of equal mass binaries [102, 103]. Nevertheless, it has been shown in Refs. [102, 104] that `IMRPhenomPv2_NRTidalv1` and `IMRPhenomPv2_NRTidalv2` models give a good approximation of waveforms even for the unequal mass binaries by computing the mismatch. Therefore, we do not expect the NR tuning of tidal effects using equal mass BNSs to impact our analysis.

[6] It should be noted that if we consider the waveform models like `TEOBResumS` [84, 112] which predict higher tidal deformabilities [7], we expect that soft EOSs would not be supported, implying a support for larger NS radius. This is similar to the results obtained in Sect. 7.4. However, a quantitative estimate of the impact of this model is left for future work.

The waveform model used by Agathos et al. for the BNS injections is TEOBResumS [84, 112] with the non-spinning tidal model of Ref. [113] and source parameters matching those of the NR simulations, as NR data do not span sufficiently many inspiral cycles to be used for injections. The Bayesian data analysis to recover the source parameters is performed using TaylorF2 and IMRPhenomPv2_NRTidalv1, restricted to the low-spin prior of Ref. [6]. However, as mentioned in Agathos et al., due to an error in the NRTidal version used, the IMRPhenomPv2_NRTidalv1 analysis led to biases in the tidal deformability inference; we therefore perform our analysis only on their TaylorF2 results.

7.4 Results

In this Section we discuss our results for (i) the analysis for the GW170817 event and (ii) the analysis for the injection studies of Agathos et al.

7.4.1 GW170817 Analysis

The analysis considered in this work to improve the inference of properties of NSs by considering the GW170817 signal as a BNS merger,[7] is orthogonal to assuming it as an NS-BH event to infer NS properties [114]. As mentioned previously for BNS mergers, we can classify the merger remnant as either a prompt collapse to a BH or a non-prompt collapse. Due to the observations of EM counterparts for the GW170817 signal, we consider the hypothesis of it to be a non-prompt collapse event for our analysis and apply our method of Sect. 7.3.3 at the post-processing level, to the posteriors of the spectral EOS parameters for the two analyses of the GW170817 signal performed in Ref. [98]:

- without maximum mass constraint on $M_{\max}^{\mathrm{TOV}}$
- with the hard constraint on the maximum mass, that is $M_{\max}^{\mathrm{TOV}} > 1.97\ \mathrm{M}_{\odot}$, corresponding to the observation of the binary pulsar PSR J0348+0432 [50]

We use the two sets of q-dependent M_{thr} fits of Bauswein+ and Kölsch+. For the Kölsch+ fits, however, we only present the analysis for the maximum mass constraint data sample, since the best M_{thr} fit given in Kölsch et al. is based on the NS radius at $1.6\ \mathrm{M}_{\odot}$, i.e. $R_{1.6\ \mathrm{M}_{\odot}}$.[8]

[7] Additionally, it has been shown in Ref. [114] using both GW and EM observations that only $<$ 40% of the binary parameters that are consistent with the GW observations yield EM counterparts if GW170817 was an NSBH event. Hence, the GW170817 signal can be more reliably considered as a BNS merger with non-prompt collapse hypothesis.

[8] That fiducial NS mass is larger than the maximum mass supported by some samples in the unconstrained dataset. Nevertheless, we have confirmed that after truncating these samples, the resulting posteriors agree well with those derived from the constrained dataset.

We compare the non-prompt collapse subset of the binary parameters like radii (r_1, r_2), (dimensionless) tidal deformability (Λ_1, Λ_2), the mass-ratio (q), and component masses (m_1, m_2) evaluated using the $M_{\rm thr}$ fits with the corresponding parameters of the complete LVC data sample. The results of our analysis for GW170817 posteriors for both the Bauswein+ and Kölsch+ fits are shown in Fig. 7.1 and Fig. 7.2, respectively.

Within the 90% credible limit, we obtain $m_1 \in (1.37, 1.61)\mathrm{M}_\odot$, $m_2 \in (1.16, 1.35)\mathrm{M}_\odot$ for the unconstrained dataset, and $m_1 \in (1.37, 1.62)\mathrm{M}_\odot$, $m_2 \in (1.16, 1.35)\mathrm{M}_\odot$ for the constrained dataset, consistent with the results of Ref. [6]. After removing the prompt collapse cases (i.e., removing 64.80% and 18.49% of the data sample from the datasets without constraint and with constraints, respectively) using Bauswein+ fits from the data sample, we obtain the component masses for the remaining non-prompt collapse data sample as $m_1 \in (1.37, 1.58)\mathrm{M}_\odot, m_2 \in (1.18, 1.36)\mathrm{M}_\odot$ for the unconstrained dataset, and $m_1 \in (1.37, 1.60)\mathrm{M}_\odot$, $m_2 \in (1.17, 1.35)\mathrm{M}_\odot$ for the constrained dataset, which within the 90% credible limit are also consistent with the results of Ref. [6]. Likewise, after removing the prompt collapse cases (i.e., removing 14.61% of the data sample) using the Kölsch+ fit from the constrained dataset, we obtain the component masses for the remaining non-prompt collapse data sample as $m_1 \in (1.37, 1.59)\mathrm{M}_\odot$ and $m_2 \in (1.17, 1.35)\mathrm{M}_\odot$, also consistent with the results of Ref. [6] within the 90% credible limit.

For the radii of the two NSs, within the 90% credible limit, we obtain $r_1 = 11.2^{+1.8}_{-1.5}$ km, $r_2 = 11.5^{+1.6}_{-1.3}$ km for the data sample without maximum mass constraint, and $r_1 = 11.9^{+1.4}_{-1.4}$ km, $r_2 = 11.9^{+1.4}_{-1.3}$ km for the data sample with maximum mass constraint for the LVC collaboration dataset [32].

Thus by taking into account the respective uncertainties in the radii, we see that the maximum mass constraint rules out the possibility of radii < 10 km compared to the data sample without maximum mass constraint. This is due to the fact that the soft EOSs that predict low NS radii do not support massive NSs. For the non-prompt collapse data sample based on the Bauswein+ fit, we obtain $r_1 = 12.3^{+1.2}_{-1.3}$ km, $r_2 = 12.3^{+1.2}_{-1.4}$ km for the unconstrained dataset and $r_1 = 12.1^{+1.3}_{-1.2}$ km, $r_2 = 12.1^{+1.2}_{-1.2}$ km for the constrained dataset. This shows that under the non-prompt collapse hypothesis, based on the q-dependent Bauswein+ fit, our GW analysis alone already rules out the possibility of NS radii < 10 km, even without using the observations of heavy NSs.

Based on the Kölsch+ fit, the radii of the two NSs within 90% credible limit for the non-prompt collapse dataset are now $r_1 = 12.1^{+1.2}_{-1.2}$ km and $r_2 = 12.1^{+1.2}_{-1.3}$ km. The Kölsch+ fit also rules out radii < 10 km consistent with the findings of Refs. [6, 32] for the constrained dataset, as well as with the $R_{1.6\mathrm{M}_\odot}$ bound of Ref. [29]. Additionally, we observe an improvement in the radii constraints as the uncertainty of NS radii decreases for the non-prompt collapse data samples. The uncertainty in NS radii for the unconstrained LVC dataset decreases from 3.3 km to 2.6 km after removing the prompt collapse cases using the Bauswein+ fit. Similarly, the uncertainty in NS radii for the constrained LVC dataset decreases from 2.8 km to 2.5 km for Bauswein+ and Kölsch+ fits, based on non-prompt collapse data samples. This

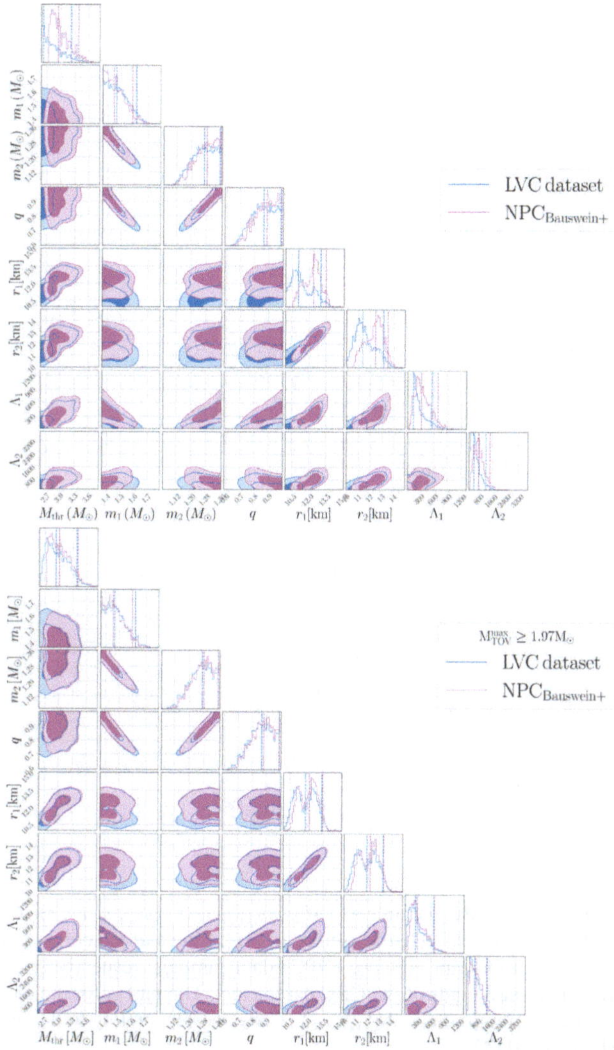

Fig. 7.1 Corner plot for the CBC parameters ($M_{\text{thr}}, m_1, m_2, \Lambda_1, \Lambda_2, q, r_1, r_2$) of the GW signal GW1701817 based on the Bauswein+ fit of Bauswein et al. [30]. Blue and red colours indicate the LVC and the corresponding non-prompt collapse (NPC) data samples, respectively. The dotted lines show the 90% credible limit bounds of the CBC parameters. The upper plot corresponds to the data sample without the maximum mass constraint and the lower plot corresponds to the data sample with a hard constraint on the maximum mass, $M_{\text{max}}^{\text{TOV}} > 1.97 M_\odot$, corresponding to the observation of the binary pulsar PSR J0348+0432, respectively

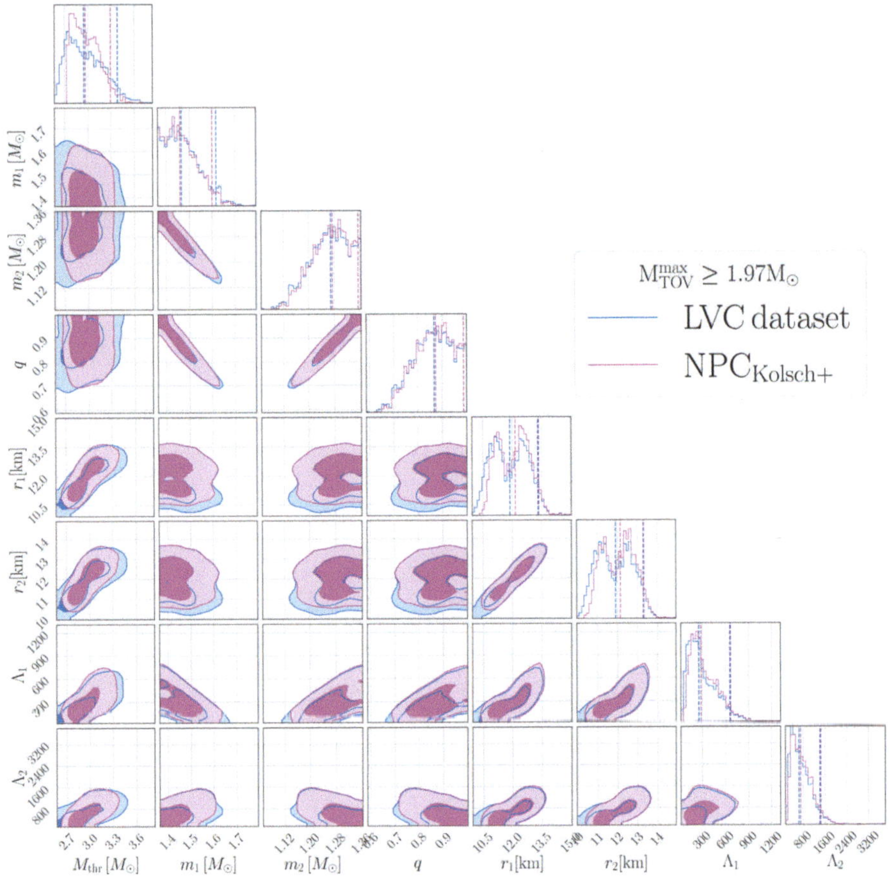

Fig. 7.2 Corner plot for the CBC parameters (M_{thr}, m_1, m_2, Λ_1, Λ_2, q, r_1, r_2) of the GW signal GW1701817 based on the Kölsch+ fit of Kölsch et al. [31]. Blue and red colours indicate the LVC and the corresponding non-prompt collapse (NPC) data samples, respectively. The dotted lines show the 90% credible limit bounds of the CBC parameters. The plot is for the data sample with a hard constraint on the maximum mass, $M_{\rm max}^{\rm TOV} > 1.97 M_{\odot}$, corresponding to the observation of the binary pulsar PSR J0348+0432

improvement in constraints on the radii implies improved constraints on the EOSs of the NS.

To illustrate the impact of our analysis, we plot the mass-radius curve for the posterior of the GW170817 signal along with the mass-radius curve for selected EOSs (WFF1, APR4, SLy, MPA1, H4) in Figs. 7.3a and 7.3b. Within the 90% credible limit, the EOSs consistent with the non-prompt collapse dataset are only APR4, SLy, and MPA1, compared to the complete LVC dataset that also supports WFF1.

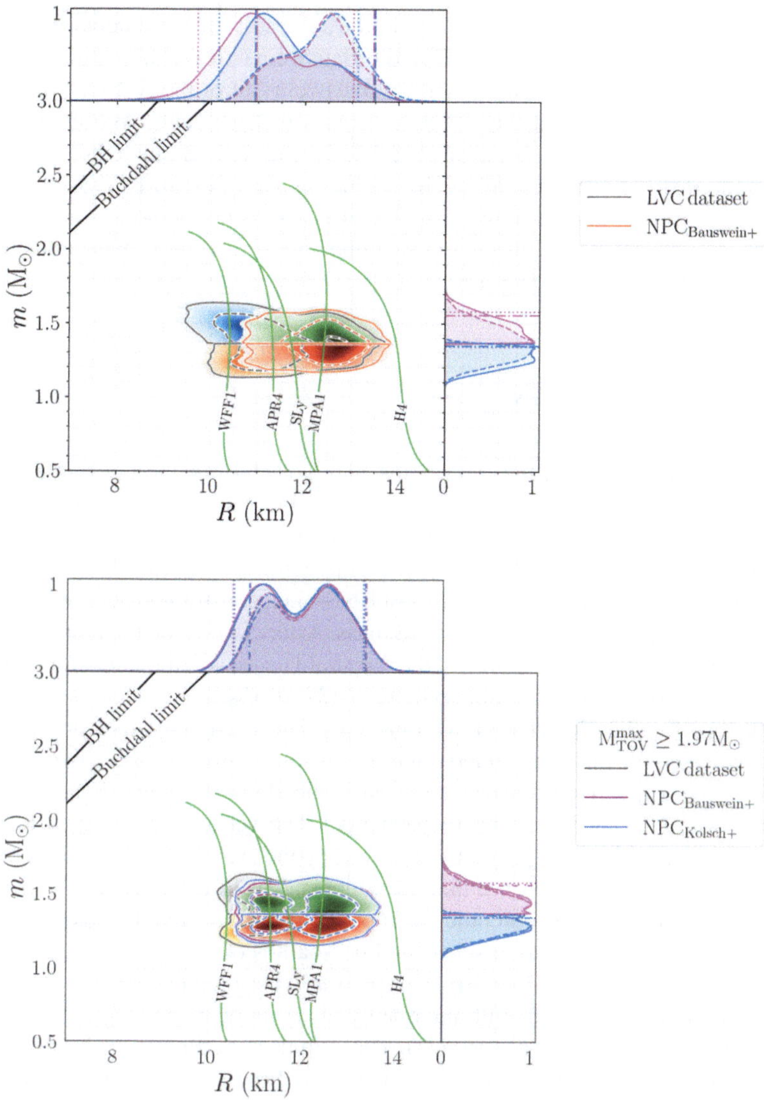

Fig. 7.3 Mass-Radius posterior plots for the mass m and radius R of each binary component of the GW event GW170817 for both the data sample without the maximum mass constraint in the upper panel, and in the lower panel, the data sample with the hard constraint $M_{\mathrm{max}}^{\mathrm{TOV}} > 1.97 M_\odot$, corresponding to the observation of the binary pulsar PSR J0348+0432. The black solid lines in the top left indicate the Schwarzschild limit ($R = 2m$), and the Buchdahl limit ($R = \frac{9}{4}m$). Mass-radius curves of selected EOSs are plotted in green colour. In the one-dimensional plots, solid and dotted lines indicate the LVC data sample and its 90% credible limit bounds, respectively, and dashed and dash-dotted lines indicate the corresponding non-prompt collapse (NPC) data sample and its 90% credible limit bounds, respectively. The contour plots with solid lines indicate the 90% credible limit and the dashed lines indicate the 50% credible limit with primary and the secondary component of the binary denoted by the top and the bottom half of the contour, respectively

This follows from the improvement in the NS radii measurements, both in the case of the unconstrained dataset using the Bauswein+ fit for $M_{\rm thr}$, and in the case of the constrained dataset using either the Bauswein+ fit or the Kölsch+ fit.

Overall, we observe that the effect of using a q-dependent classification of the remnant improves our constraints on the NS EOS and radii in much the same way as a maximum mass constraint does, i.e. by excluding part of the soft-EOS corner of the parameter space. In this particular case, the effect of using the non-prompt collapse hypothesis is stronger than the effect of imposing a maximum mass constraint at 1.97 $M_\odot$, as highlighted in Fig. 7.3.

7.4.2 Injection Studies

To check the robustness of the Bauswein+ and Kölsch+ fits, we use progenitor and remnant information from the existing NR simulations of BNS mergers given in Agathos et al.. We generate posteriors of CBC parameters using the TaylorF2 waveform model for each of these BNS systems. Similar to the GW1701817 analysis of Sect. 7.4.1, we consider the two cases of BNSs, i.e., constrained and unconstrained datasets with respect to the maximum NS mass. After classifying the merger remnant of each dataset using the above fits for the threshold mass, we calculate the cumulative probability distribution of prompt collapse, $P(M > M_{\rm thr})$.

Here too, for the Kölsch+ fit we only calculate the cumulative probability of prompt collapse for the constrained dataset. The cumulative probability distribution (CDF) plots for the constrained dataset using the Bauswein+ and Kölsch+ fits are shown in Fig. 7.4a and Fig. 7.4b, respectively.[9] The vertical line in the plot corresponds to the threshold of prompt collapse (beyond which the total mass exceeds the threshold mass). We observe that the q-dependent Bauswein+ and Kölsch+ fits correctly infer the merger outcome for most of the BNS models except for a few cases, all of which were also misclassified in Agathos et al..

Then, we compare the probability of prompt collapse predicted by the q-dependent Bauswein+ and Kölsch+ fits with the one by the q-independent $M_{\rm thr}$ fit of Agathos et al. (see, Eq. (7.6) for the fit formula). The summary of the results based on these fit formulae along with the binary parameters and the outcome of the merger for each BNS is shown in Table 7.1. We observe that both Bauswein+ and Kölsch+ fits better capture the formation of prompt collapse to a BH as the probability for prompt collapse increases for both the cases of data samples compared to q-independent $M_{\rm thr}$ fit of Agathos et al.. Furthermore, we observe that compared to Bauswein+, the Kölsch+ fit is better at predicting the type of the merger remnant, as it typically gives higher probabilities when a prompt collapse does occur and lower probabilities in the cases resulting in a (meta-) stable remnant. We see that even for the few cases where the merger remnant is misclassified (SFHo $1.4 + 1.4$, ALF2 $1.5 + 1.5$), with respect

---

[9] For the unconstrained dataset CDF plots using Bauswein+ fit, see Fig. 7.5 in Sect. 7.6.

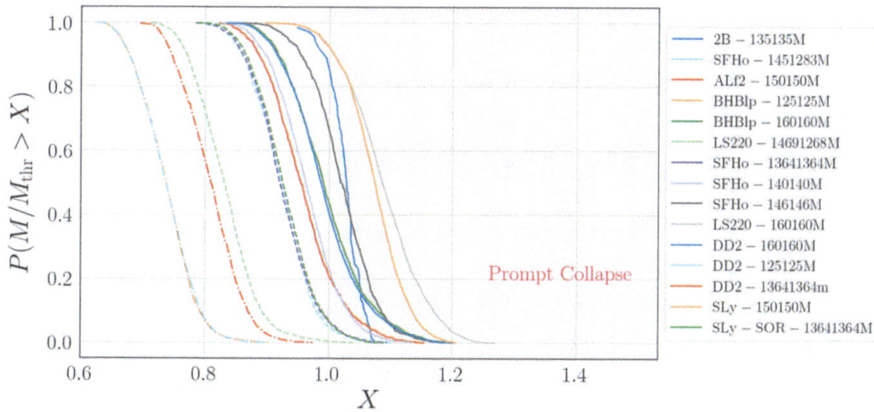

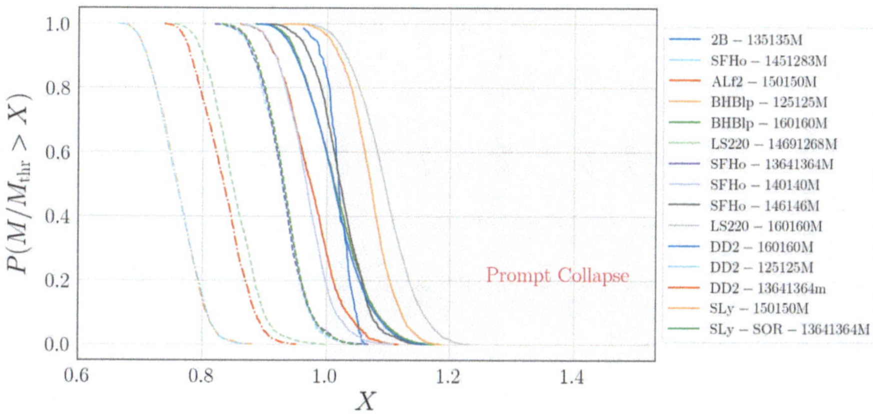

Fig. 7.4 Cumulative probability of $M/M_{\rm thr}$, the ratio between the total mass M and the threshold mass $M_{\rm thr}$, for the injection study using simulated BNS systems of Table 7.1 with the hard constraint on the maximum mass, $M_{\rm max}^{\rm TOV} > 1.97 M_{\odot}$, corresponding to the observation of the binary pulsar PSR J0348+0432. The probability of prompt collapse for each binary neutron star system is at $X = 1$. The solid, dashed and dashed-dotted lines correspond to the formation of a BH, Hyper-massive NS (HMNS), and Massive NS (MNS), respectively. The plots correspond to the Bauswein+ threshold mass fit of Eq. (7.3) in the upper panel, and the Kölsch+ threshold mass fit of Eq. (7.4) in the lower panel, respectively

to the results of Agathos et al., the probability of prompt collapse has significantly moved in the correct direction.

Finally, as mentioned in Sect. 7.3.1, we also use the $M_{\rm thr}^{K1}$ fit dependent on the estimated value of the threshold mass at ($q = 1$) from NR simulation data $\left(M_{\rm thr}^{q=1}\right)$ together with $M_{\rm max}^{\rm TOV}$ and q for our analysis (see Eq. (7.5) for the fitting formula). The values of the threshold mass at ($q = 1$) for the EOSs of interest are obtained from the NR data of Refs. [31, 32, 51, 52] (see, Table 7.2). The probability of prompt collapse computed corresponding to this fit is also given in Table 7.1. From Table 7.1, we see

Table 7.1 Summary of probability of prompt collapse (%) for the known inspiral-merger waveforms of Agathos et al. [32]. The collapse time $t_{\rm BH}$ is time measured from the peak of GW amplitude to the first formation of the apparent horizon, called the merger time. The threshold mass fit formulas $M_{\rm thr}(M_{\rm max}^{\rm TOV}, C_{\rm max})$, $M_{\rm thr}(M_{\rm max}^{\rm TOV}, R_{\rm max}, q)$, $M_{\rm thr}(M_{\rm max}, R_{1.6{\rm M}_\odot}, q)$, and $M_{\rm thr}(M_{\rm max}^{\rm TOV}, M_{\rm thr,(q=1)}, q)$ are given in Eqs. (7.3)–(7.6), respectively

EOS	$m_1 ({\rm M}_\odot)$	$m_2 ({\rm M}_\odot)$	$t_{\rm BH}[ms]$	Remnant at $t \sim 3ms$	$M_{\rm thr}({M_{\rm max}}^{\rm TOV}, C_{\rm max})$ Agathos et al. [32]		$M_{\rm thr}(M_{\rm max}^{\rm TOV}, R_{\rm max}, q)$ Bauswein et al. [30]		$M_{\rm thr}(M_{\rm max}^{\rm TOV}, R_{1.6{\rm M}_\odot}, q)$ Kolsch et al. [31]	$M_{\rm thr}(M_{\rm max}^{\rm TOV}, M_{\rm thr}^{q=1}, q)$ Kolsch et al. [31]
					$P_{\rm PC}^{\rm M_{thr}}$	$P_{\rm PC}^{\rm M_{thr}, M_{max}^{TOV}}$	$P_{\rm PC}^{\rm M_{thr}}$	$P_{\rm PC}^{\rm M_{thr}, M_{max}^{TOV}}$	$P_{\rm PC}^{\rm M_{thr}, M_{max}^{TOV}}$	$P_{\rm PC}^{\rm M_{thr}, M_{max}^{TOV}}$
2B	1.35	1.35	0.49	BH	99.5	67.1	99.9	86.3	83.6	100.0
SFHo	1.40	1.40	1.07	BH	46.7	8.0	58.8	23.6	18.8	17.6
SFHo	1.46	1.46	0.70	BH	77.2	48.1	85.6	66.5	71.5	100.0
ALF2	1.50	1.50	0.64	BH	25.7	5.7	36.6	21.1	28.6	84.0
BHBΛϕ	1.60	1.60	0.99	BH	40.7	27.2	53.4	44.0	60.1	100.0
LS220	1.60	1.60	0.63	BH	92.4	88.1	95.3	92.6	97.9	100.0
DD2	1.60	1.60	~3	BH	37.7	24.2	50.2	40.9	59.7	32.0
SLy	1.50	1.50	0.99	BH	96.1	87.9	97.6	92.5	96.2	100.0
SFHo	1.364	1.364	~4	HMNS	25.7	0.5	34.1	6.7	4.0	3.4
SLy-SOR	1.364	1.364	~14	HMNS	29.8	0.5	39.1	7.2	3.4	4.8
SFHo	1.45	1.283	~12	HMNS	25.9	0.4	34.7	5.0	2.7	2.9
LS220	1.469	1.268	~33	HMNS	0.8	0.0	1.9	0.4	0.0	0.1
BHBΛϕ	1.25	1.25	>20	MNS	0.0	0.0	0.0	0.0	0.0	0.0
DD2	1.25	1.25	>20	MNS	0.0	0.0	0.0	0.0	0.0	0.0
DD2	1.364	1.364	~21	MNS	0.1	0.0	0.9	0.0	0.0	0.0

Table 7.2 NR simulation based values of $M_{\text{thr}}^{q=1}$ for EOSs considered in Agathos et al. [32]

EOS	$M_{\text{thr}}^{q=1}$	Reference
2B	2.43	[32]
SFHo	2.824	[51, 52]
ALF2	2.963	[31]
BHBΛΦ	3.024	[51, 52]
LS220	2.956	[51, 52]
DD2	3.274	[51, 52]
SLy	2.756	[31]

that this fit better captures both the formation of a prompt collapse to a BH and a non-prompt collapse compared to all other M_{thr} fits considered in our analysis for most of the BNS systems; however, it is only available for a small discrete set of EOSs.

7.5 Conclusions

We analyse the gravitational wave signal GW170817 using the q-dependent threshold mass fits given in Bauswein et al. [30] and Kölsch et al. [31]. For our analysis, we consider two analyses of the GW170817 signal - without the maximum mass constraint data sample and with the hard constraint on the maximum mass data sample corresponding to the observation of binary pulsar PSR J0348+0432. Using the threshold mass fits we identify the prompt collapse cases which we then remove from the two data samples for our analysis. Then we compare the non-prompt collapse data sample of each case with its corresponding complete LVC data sample. We observe that the non-prompt collapse data sample removes radii < 10 km even without imposing the maximum mass constraint on the data sample in comparison to results observed in Refs. [6, 32] where radii < 10 km are only excluded by the data sample with the maximum mass constraint. We also see that, for the non-prompt collapse data sample the uncertainty of NS radii decreases implying improved constraints on the NS radii and hence the EOSs. As an example, we observe that the EOS WFF1 is not supported by the non-prompt collapse data sample for both of the cases of GW170817 signal compared to the results of Refs. [6, 32].

The threshold mass fits are then validated with a set of 17 injections given in Agathos et al. [32]. The merger remnant for all the signals is not just correctly inferred from both the fits, but the probability of the formation of a prompt collapse is also improved compared to the q-independent fit of Agathos et al. with the exception of a few signals. Additionally, we observe that the probability of formation of both a prompt collapse and a non-prompt collapse remnant has improved for the

q-dependent threshold mass fits of Kölsch et al. compared to the fits of Bauswein et al. and Agathos et al..

Application of either of the fits, Bauswein+ or Kölsch+, to the GW170817 analysis gives similar improvements for the posterior probability distribution functions of NS masses and radii. This suggests that any differences between these two fits have a relatively small impact for GW170817-like signals, which should also allow us to improve the constraints on the NS properties for similar detections in the future. Nevertheless, the BNS parameter space, especially in the regions close to the prompt collapse threshold, needs to be explored more densely with future numerical relativity simulations for a variety of EOSs.

7.6 Effect of $M_{\max}^{\mathrm{TOV}}$ Constraint

In this Section, we analyse the impact of the maximum mass constraint corresponding to the observation of binary pulsar J0348+0432 on the known BNS inspiral-merger waveforms of Agathos et al. [32]. For this, we first calculate the cumulative probability of prompt collapse ($P(M > M_{\mathrm{thr}})$) for the data sample without the maximum mass constraint and then compare it with the results presented in Fig. 7.5a. The cumulative probability plot for the data sample without maximum mass constraint is shown in Fig. 7.5.

On comparing Fig. 7.5 with Fig. 7.5a, we observe that the ratio M/M_{thr} decreases for the data sample with the maximum mass constraint compared to the data sample without. This can be due to an increase in the M_{thr} values, which is expected, as the

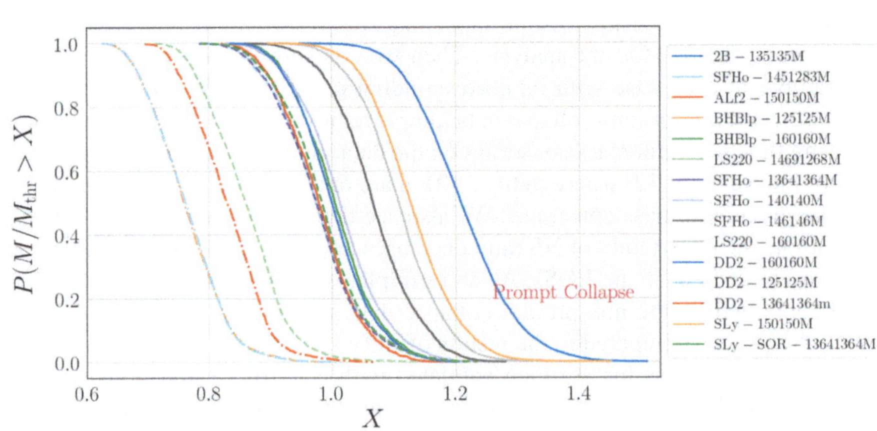

Fig. 7.5 Cumulative probability of M/M_{thr}, the ratio between the total mass M and the threshold mass M_{thr}, for the injection study using simulated BNS systems of Table 7.1 for the Bauswein+ fit of Eq. (7.3). The probability of prompt collapse for each binary neutron star system by definition is at $X = 1$. The solid, dashed and dashed-dotted lines correspond to the formation of a BH, HMNS, and MNS, respectively

maximum mass constraint removes the soft part of the EOSs. This is also reflected in the results in Table 7.1, where we see a decreasing trend in the probability of prompt collapse for the maximum mass constraint data sample compared to the data sample without maximum mass constraint.

References

1. Jain T, Agathos M (2024) Improving inference on neutron star properties using information from binary merger remnants. arXiv:2404.12126 [gr-qc]
2. Abbott BP et al (2016) (LIGO Scientific, Virgo), Observation of gravitational waves from a binary black hole merger. Phys Rev Lett 116:061102. https://doi.org/10.1103/PhysRevLett.116.061102, arXiv:1602.03837 [gr-qc]
3. Aasi J et al (2015) (LIGO Scientific), Advanced LIGO, Class Quant Grav 32:074001. https://doi.org/10.1088/0264-9381/32/7/074001, arXiv:1411.4547 [gr-qc]
4. Acernese F et al (2015) (VIRGO), Advanced Virgo: a second-generation interferometric gravitational wave detector. Class Quant Grav 32:024001. https://doi.org/10.1088/0264-9381/32/2/024001, arXiv:1408.3978 [gr-qc]
5. Abbott BP et al (2017) (LIGO Scientific, Virgo), GW170817: observation of gravitational waves from a binary neutron star inspiral. Phys Rev Lett 119:161101 (2017). https://doi.org/10.1103/PhysRevLett.119.161101, arXiv:1710.05832 [gr-qc]
6. Abbott BP et al (2019) (LIGO Scientific, Virgo), Properties of the binary neutron star merger GW170817. Phys Rev X 9:011001. https://doi.org/10.1103/PhysRevX.9.011001 , arXiv:1805.11579 [gr-qc]
7. Abbott BP et al (2019) (LIGO Scientific, Virgo), GWTC-1: a gravitational-wave transient catalog of compact binary mergers observed by ligo and virgo during the first and second observing runs. Phys Rev X 9:031040. https://doi.org/10.1103/PhysRevX.9.031040 , arXiv:1811.12907 [astro-ph.HE]
8. Abbott BP et al (2020) (LIGO Scientific, Virgo), GW190425: observation of a compact binary coalescence with total mass $\sim 3.4 M_{\odot}$. Astrophys J Lett 892:L3. https://doi.org/10.3847/2041-8213/ab75f5 , arXiv:2001.01761 [astro-ph.HE]
9. Shibata M (2005) Constraining nuclear equations of state using gravitational waves from hypermassive neutron stars. Phys Rev Lett 94:201101. https://doi.org/10.1103/PhysRevLett.94.201101, arXiv:gr-qc/0504082 <error l="308" c="Invalid command: paragraph not started." />
10. Shibata M, Taniguchi K (2006) Merger of binary neutron stars to a black hole: disk mass, short gamma-ray bursts, and quasinormal mode ringing. Phys Rev D 73:064027. https://doi.org/10.1103/PhysRevD.73.064027, arXiv:astro-ph/0603145
11. Baiotti L, Giacomazzo B, Rezzolla L (2008) Accurate evolutions of inspiralling neutron-star binaries: prompt and delayed collapse to black hole. Phys Rev D 78:084033. https://doi.org/10.1103/PhysRevD.78.084033, arXiv:0804.0594 [gr-qc]
12. Hotokezaka K, Kyutoku K, Okawa H, Shibata M, Kiuchi K (2011) Binary neutron star mergers: dependence on the nuclear equation of state. Phys Rev D 83:124008. https://doi.org/10.1103/PhysRevD.83.124008 , arXiv:1105.4370 [astro-ph.HE]
13. Bauswein A, Baumgarte TW, Janka HT (2013) Prompt merger collapse and the maximum mass of neutron stars. Phys Rev Lett 111:131101. https://doi.org/10.1103/PhysRevLett.111.131101 , arXiv:1307.5191 [astro-ph.SR]
14. Bauswein A, Stergioulas N (2017) Semi-analytic derivation of the threshold mass for prompt collapse in binary neutron star mergers. Mon Not Roy Astron Soc 471:4956–4965. https://doi.org/10.1093/mnras/stx1983, arXiv:1702.02567 [astro-ph.HE]
15. Bernuzzi S (2020) Neutron star merger remnants. Gen Rel Grav 52:108. https://doi.org/10.1007/s10714-020-02752-5, arXiv:2004.06419 [astro-ph.HE]

16. Abbott BP et al (2017) (LIGO Scientific, Virgo, Fermi-GBM, INTEGRAL), Gravitational waves and gamma-rays from a binary neutron star merger: GW170817 and GRB 170817A. Astrophys J Lett 848:L13.). https://doi.org/10.3847/2041-8213/aa920c , arXiv:1710.05834 [astro-ph.HE]
17. Troja E et al (2017) The X-ray counterpart to the gravitational wave event GW 170817. Nature 551:71–74. https://doi.org/10.1038/nature24290, arXiv:1710.05433 [astro-ph.HE]
18. Abbott et al BP (2017) (LIGO Scientific, Virgo, Fermi GBM, INTEGRAL, IceCube, AstroSat Cadmium Zinc Telluride Imager Team, IPN, Insight-Hxmt, ANTARES, Swift, AGILE Team, 1M2H Team, Dark Energy Camera GW-EM, DES, DLT40, GRAWITA, Fermi-LAT, ATCA, ASKAP, Las Cumbres Observatory Group, OzGrav, DWF (Deeper Wider Faster Program), AST3, CAASTRO, VINROUGE, MASTER, J-GEM, GROWTH, JAGWAR, CaltechNRAO, TTU-NRAO, NuSTAR, Pan-STARRS, MAXI Team, TZAC Consortium, KU, Nordic Optical Telescope, ePESSTO, GROND, Texas Tech University, SALT Group, TOROS, BOOTES, MWA, CALET, IKI-GW Follow-up, H.E.S.S., LOFAR, LWA, HAWC, Pierre Auger, ALMA, Euro VLBI Team, Pi of Sky, Chandra Team at McGill University, DFN, ATLAS Telescopes, High Time Resolution Universe Survey, RIMAS, RATIR, SKA South Africa/MeerKAT), Multi-messenger Observations of a Binary Neutron Star Merger. Astrophys J Lett 848:L12. https://doi.org/10.3847/2041-8213/aa91c9, arXiv:1710.05833 [astro-ph.HE]
19. Tanaka M et al (2017) Kilonova from post-merger ejecta as an optical and near-infrared counterpart of GW170817. Publ Astron Soc Jap 69:psx12. https://doi.org/10.1093/pasj/psx121 , arXiv:1710.05850 [astro-ph.HE]
20. Tanvir NR et al (2017) The emergence of a lanthanide-rich kilonova following the merger of two neutron stars. Astrophys J Lett 848:L27. https://doi.org/10.3847/2041-8213/aa90b6, arXiv:1710.05455 [astro-ph.HE]
21. Cowperthwaite PS et al (2017) The electromagnetic counterpart of the binary neutron star merger LIGO/Virgo GW170817. II. UV, Optical, and near-infrared light curves and comparison to kilonova models. Astrophys J Lett 848:L17. https://doi.org/10.3847/2041-8213/aa8fc7, arXiv:1710.05840 [astro-ph.HE]
22. Nicholl M et al (2017) The electromagnetic counterpart of the binary neutron star merger LIGO/VIRGO GW170817. III. Optical and UV spectra of a blue kilonova from fast polar ejecta. Astrophys J Lett 848:L18. https://doi.org/10.3847/2041-8213/aa9029 , arXiv:1710.05456 [astro-ph.HE]
23. Chornock R et al (2017) The electromagnetic counterpart of the binary neutron star merger LIGO/VIRGO GW170817. IV. Detection of near-infrared signatures of r-process nucleosynthesis with Gemini-South. Astrophys J Lett 848:L19. https://doi.org/10.3847/2041-8213/aa905c , arXiv:1710.05454 [astro-ph.HE]
24. Coulter DA et al (2017) Swope Supernova Survey 2017a (SSS17a), the optical counterpart to a gravitational wave source. Science 358:1556. https://doi.org/10.1126/science.aap9811, arXiv:1710.05452 [astro-ph.HE]
25. Shibata M, Uryu K (2000) Simulation of merging binary neutron stars in full general relativity: Gamma = two case. Phys Rev D 61:064001. https://doi.org/10.1103/PhysRevD.61.064001 , arXiv:gr-qc/9911058
26. Shibata M, Uryu K (2002) Gravitational waves from the merger of binary neutron stars in a fully general relativistic simulation. Prog Theor Phys 107:265. https://doi.org/10.1143/PTP.107.265, arXiv:gr-qc/0203037
27. Anderson M, Hirschmann EW, Lehner L, Liebling SL, Motl PM, Neilsen D, Palenzuela C, Tohline JE (2008) Simulating binary neutron stars: dynamics and gravitational waves. Phys Rev D 77:024006. https://doi.org/10.1103/PhysRevD.77.024006, arXiv:0708.2720 [gr-qc]
28. Bauswein A, Blacker S, Vijayan V, Stergioulas N, Chatziioannou K, Clark JA, Bastian N-UF, Blaschke DB, Cierniak M, Fischer T (2020) Equation of state constraints from the threshold binary mass for prompt collapse of neutron star mergers. Phys Rev Lett 125:141103. https://doi.org/10.1103/PhysRevLett.125.141103 , arXiv:2004.00846 [astro-ph.HE]
29. Bauswein A, Just O, Janka H-T, Stergioulas N (2017) Neutron-star radius constraints from GW170817 and future detections. Astrophys J Lett 850:L34. https://doi.org/10.3847/2041-8213/aa9994, arXiv:1710.06843 [astro-ph.HE]

30. Bauswein A, Blacker S, Lioutas G, Soultanis T, Vijayan V, Stergioulas N (2021) Systematics of prompt black-hole formation in neutron star mergers. Phys Rev D 103:123004. https://doi.org/10.1103/PhysRevD.103.123004, arXiv:2010.04461 [astro-ph.HE]
31. Kölsch M, Dietrich T, Ujevic M, Bruegmann B (2021) Investigating the mass-ratio dependence of the prompt-collapse threshold with numerical-relativity simulations, arXiv:2112.11851 [gr-qc]
32. Agathos M, Zappa F, Bernuzzi S, Perego A, Breschi M, Radice D (2020) Inferring prompt black-hole formation in neutron star mergers from gravitational-wave data. Phys Rev D 101:044006. https://doi.org/10.1103/PhysRevD.101.044006, arXiv:1908.05442 [gr-qc]
33. Bernuzzi S, Hilditch D (2010) Constraint violation in free evolution schemes: comparing BSSNOK with a conformal decomposition of Z4. Phys Rev D 81:084003. https://doi.org/10.1103/PhysRevD.81.084003, arXiv:0912.2920 [gr-qc]
34. Hilditch D, Bernuzzi S, Thierfelder M, Cao Z, Tichy W, Bruegmann B (2013) Compact binary evolutions with the Z4c formulation. Phys Rev D 88:084057. https://doi.org/10.1103/PhysRevD.88.084057, arXiv:1212.2901 [gr-qc]
35. van Meter JR, Baker JG, Koppitz M, Choi D-I (2006) How to move a black hole without excision: Gauge conditions for the numerical evolution of a moving puncture. Phys Rev D 73. https://doi.org/10.1103/physrevd.73.124011
36. Alcubierre M, Bruegmann B, Diener P, Koppitz M, Pollney D, Seidel E, Takahashi R (2003) Gauge conditions for long term numerical black hole evolutions without excision. Phys Rev D 67:084023. https://doi.org/10.1103/PhysRevD.67.084023. arXiv:gr-qc/0206072
37. Bona C, MassA A3 J, Stela J, Seidel E (1997) A class of hyperbolic gauge conditions. In: Jantzen RT, Keiser GC, Ruffini R (eds) The seventh marcel grossmann meeting: on recent developments in theoretical and experimental general relativity, gravitation, and relativistic field theories. World Scientific. https://doi.org/10.1142/2569
38. Oechslin R, Rosswog S, Thielemann FK (2002) Conformally flat smoothed particle hydrodynamics: application to neutron star mergers. Phys Rev D 65:103005. https://doi.org/10.1103/PhysRevD.65.103005. arXiv:gr-qc/0111005
39. Oechslin R, Janka HT, Marek A (2007) Relativistic neutron star merger simulations with non-zero temperature equations of state. 1. Variation of binary parameters and equation of state. Astron Astrophys 467:395. https://doi.org/10.1051/0004-6361:20066682. arXiv:astro-ph/0611047
40. Bauswein A, Janka HT, Oechslin R (2010) Testing approximations of thermal effects in neutron star merger simulations. Phys Rev D 82:084043. https://doi.org/10.1103/PhysRevD.82.084043. arXiv:1006.3315 [astro-ph.SR]
41. Isenberg J, Nester J (1980) General relativity and gravitation, p 23. Plenum Press, New York
42. Wilson JR, Mathews GJ, Marronetti P (1996) Relativistic numerical model for close neutron star binaries. Phys Rev D 54:1317–1331. https://doi.org/10.1103/PhysRevD.54.1317. arXiv:gr-qc/9601017
43. Monaghan JJ (1992) Smoothed particle hydrodynamics. Ann Rev Astron Astrophys 30:543–574. https://doi.org/10.1146/annurev.aa.30.090192.002551
44. Hockney R, Eastwood J (1988) Computer simulation using particles, 1st edn. CRC Press. https://doi.org/10.1201/9780367806934
45. Berger MJ, Oliger J (1984) Adaptive mesh refinement for hyperbolic partial differential equations. J Comput Phys 53:484–512. https://doi.org/10.1016/0021-9991(84)90073-1
46. Hess R (1999) Dynamically Adaptive Multigrid on Parallel Computers for a Semi-Implicit Discretization of the Shallow Water Equations (GMD Forschungszentrum Informationstechnik GmbH. St, Augustin, Germany
47. Alford M, Braby M, Paris MW, Reddy S (2005) Hybrid stars that masquerade as neutron stars. Astrophys J 629:969–978. https://doi.org/10.1086/430902. arXiv:nucl-th/0411016
48. Douchin F, Haensel P (2001) A unified equation of state of dense matter and neutron star structure. Astron Astrophys 380:151. https://doi.org/10.1051/0004-6361:20011402. arXiv:astro-ph/0111092

49. Lackey BD, Nayyar M, Owen BJ (2006) Observational constraints on hyperons in neutron stars. Phys Rev D 73:024021. https://doi.org/10.1103/PhysRevD.73.024021. arXiv:astro-ph/0507312
50. Antoniadis J et al (2013) A massive pulsar in a compact relativistic binary. Science 340:6131. https://doi.org/10.1126/science.1233232. arXiv:1304.6875 [astro-ph.HE]
51. Perego A, Logoteta D, Radice D, Bernuzzi S, Kashyap R, Das A, Padamata S, Prakash A (2021) Probing the incompressibility of nuclear matter at ultra-high density through the prompt collapse of asymmetric neutron star binaries. arXiv:2112.05864 [astro-ph.HE]
52. Kashyap R et al (2021) Numerical relativity simulations of prompt collapse mergers: threshold mass and phenomenological constraints on neutron star properties after GW170817. arXiv:2111.05183 [astro-ph.HE]
53. Banik S, Hempel M, Bandyopadhyay D (2014) New hyperon equations of state for supernovae and neutron stars in density-dependent hadron field theory. Astrophys J Suppl 214:22. https://doi.org/10.1088/0067-0049/214/2/22. arXiv:1404.6173 [astro-ph.HE]
54. Fortin M, Oertel M, Provid A Ancia C (2018) Hyperons in hot dense matter: what do the constraints tell us for equation of state? Publ Astron Soc Aust 35:e044
55. Marques M, Oertel M, Hempel M, Novak J (2017) New temperature dependent hyperonic equation of state: application to rotating neutron star models and I-Q relations. Phys Rev C 96:045806. https://doi.org/10.1103/PhysRevC.96.045806. arXiv:1706.02913 [nucl-th]
56. Hempel M, Schaffner-Bielich J (2010) Statistical model for a complete supernova equation of state. Nucl Phys A 837:210–254. https://doi.org/10.1016/j.nuclphysa.2010.02.010. arXiv:0911.4073 [nucl-th]
57. Typel S, Ropke G, Klahn T, Blaschke D, Wolter HH (2010) Composition and thermodynamics of nuclear matter with light clusters. Phys Rev C 81:015803. https://doi.org/10.1103/PhysRevC.81.015803. arXiv:0908.2344 [nucl-th]
58. Typel S (2005) Relativistic model for nuclear matter and atomic nuclei with momentum-dependent self-energies. Phys Rev C 71:064301. https://doi.org/10.1103/PhysRevC.71.064301 arXiv:nucl-th/0501056
59. Alvarez-Castillo D, Ayriyan A, Benic S, Blaschke D, Grigorian H, Typel S (2016) New class of hybrid EoS and Bayesian M-R data analysis. Eur Phys J A 52:69. https://doi.org/10.1140/epja/i2016-16069-2. arXiv:1603.03457 [nucl-th]
60. Akmal A, Pandharipande VR, Ravenhall DG (1998) The equation of state of nucleon matter and neutron star structure. Phys Rev C 58:1804–1828. https://doi.org/10.1103/PhysRevC.58.1804. arXiv:nucl-th/9804027
61. Goriely S, Chamel N, Pearson JM (2010) Further explorations of Skyrme-Hartree-Fock-Bogoliubov mass formulas. XII: Stiffness and stability of neutron-star matter. Phys Rev C 82:035804. https://doi.org/10.1103/PhysRevC.82.035804. arXiv:1009.3840 [nucl-th]
62. Wiringa RB, Fiks V, Fabrocini A (1988) Equation of state for dense nucleon matter. Phys Rev C 38:1010–1037. https://doi.org/10.1103/PhysRevC.38.1010
63. Lattimer JM, Douglas Swesty F (1991) A generalized equation of state for hot dense matter. Nucl Phys A 535:331–376
64. Shen G, Horowitz CJ, Teige S (2011) A new equation of state for astrophysical simulations. Phys Rev C 83:035802. https://doi.org/10.1103/PhysRevC.83.035802. arXiv:1101.3715 [astro-ph.SR]
65. Lalazissis GA, Konig J, Ring P (1997) A new parametrization for the Lagrangian density of relativistic mean field theory. Phys Rev C 55:540–543. https://doi.org/10.1103/PhysRevC.55.540. arXiv:nucl-th/9607039
66. Hempel M, Fischer T, Schaffner-Bielich J, Liebendorfer M (2012) New equations of state in simulations of core-collapse supernovae. Astrophys J 748:70. https://doi.org/10.1088/0004-637X/748/1/70. arXiv:1108.0848 [astro-ph.HE]
67. Steiner AW, Hempel M, Fischer T (2013) Core-collapse supernova equations of state based on neutron star observations. Astrophys J 774:17. https://doi.org/10.1088/0004-637X/774/1/17. arXiv:1207.2184 [astro-ph.SR]

68. Müther H, Prakash M, Ainsworth TL (1987) The nuclear symmetry energy in relativistic Brueckner-Hartree-Fock calculations. Phys Lett B 199:469–474. https://doi.org/10.1016/0370-2693(87)91611-X
69. Engvik L, Bao G, Hjorth-Jensen M, Osnes E, Ostgaard E (1996) Asymmetric nuclear matter and neutron star properties. Astrophys J 469:794. https://doi.org/10.1086/177827. arXiv:nucl-th/9509016
70. Schneider AS, Constantinou C, Muccioli B, Prakash M (2019) Akmal-Pandharipande-Ravenhall equation of state for simulations of supernovae, neutron stars, and binary mergers. Phys Rev C 100:025803. https://doi.org/10.1103/PhysRevC.100.025803. arXiv:1901.09652 [nucl-th]
71. Read JS, Lackey BD, Owen BJ, Friedman JL (2009) Constraints on a phenomenologically parameterized neutron-star equation of state. Phys Rev D 79:124032. https://doi.org/10.1103/PhysRevD.79.124032. arXiv:0812.2163 [astro-ph]
72. Glendenning NK (1985) Neutron stars are giant hypernuclei? Astrophys J 293:470–493. https://doi.org/10.1086/163253
73. Sugahara Y, Toki H (1994) Relativistic mean field theory for unstable nuclei with non-linear sigma and omega terms. Nucl Phys A 579:557–572. https://doi.org/10.1016/0375-9474(94)90923-7
74. Toki H, Hirata D, Sugahara Y, Sumiyoshi K, Tanihata I (1995) Relativistic many body approach for unstable nuclei and supernova. Nucl Phys A 588:c357–c363. https://doi.org/10.1016/0375-9474(95)00161-S
75. Bastian N-UF, Blaschke D, Fischer T, Röpke G (2018) Towards a unified quark-hadron matter equation of state for applications in astrophysics and heavy-ion collisions. Universe 4:67. https://doi.org/10.3390/universe4060067. arXiv:1804.10178 [nucl-th]
76. Cierniak M, Klähn T, Fischer T, Bastian N-U (2018) Vector-interaction-enhanced bag model. Universe 4:30. https://doi.org/10.3390/universe4020030. arXiv:1802.03214 [nucl-th]
77. Fischer T, Bastian N-UF, Wu M-R, Baklanov P, Sorokina E, Blinnikov S, Typel S, Klähn T, Blaschke DB (2018) Quark deconfinement as a supernova explosion engine for massive blue supergiant stars. Nat Astron 2:980–986. https://doi.org/10.1038/s41550-018-0583-0. arXiv:1712.08788 [astro-ph.HE]
78. Bastian N-UF (2021) Phenomenological quark-hadron equations of state with first-order phase transitions for astrophysical applications. Phys Rev D 103:023001. https://doi.org/10.1103/PhysRevD.103.023001. arXiv:2009.10846 [nucl-th]
79. Bauswein A, Bastian N-UF, Blaschke DB, Chatziioannou K, Clark JA, Fischer T, Oertel M (2019) Identifying a first-order phase transition in neutron star mergers through gravitational waves. Phys Rev Lett 122:061102. https://doi.org/10.1103/PhysRevLett.122.061102. arXiv:1809.01116 [astro-ph.HE]
80. Kaltenborn MAR, Bastian N-UF, Blaschke DB (2017) Quark-nuclear hybrid star equation of state with excluded volume effects. Phys Rev D 96:056024. https://doi.org/10.1103/PhysRevD.96.056024. arXiv:1701.04400 [astro-ph.HE]
81. Radice D, Perego A, Bernuzzi S, Zhang B (2018) Long-lived remnants from binary neutron star mergers. Mon Not Roy Astron Soc 481:3670–3682. https://doi.org/10.1093/mnras/sty2531. arXiv:1803.10865 [astro-ph.HE]
82. T. Dietrich, D. Radice, S. Bernuzzi, F. Zappa, A. Perego, B. Brügmann, S. V. Chaurasia, R. Dudi, W. Tichy, and M. Ujevic, CoRe database of binary neutron star merger waveforms, https://doi.org/10.1088/1361-6382/aaebc0 Class. Quant. Grav. **35**, 24LT01 (2018), arXiv:1806.01625 [gr-qc]
83. Radice D, Perego A, Hotokezaka K, Fromm SA, Bernuzzi S, Roberts LF (2018) Binary neutron star mergers: mass ejection, electromagnetic counterparts and nucleosynthesis. Astrophys J 869:130. https://doi.org/10.3847/1538-4357/aaf054. arXiv:1809.11161 [astro-ph.HE]
84. Bernuzzi S, Nagar A, Dietrich T, Damour T (2015) Modeling the dynamics of tidally interacting binary neutron stars up to the merger. Phys Rev Lett 114:161103. https://doi.org/10.1103/PhysRevLett.114.161103. arXiv:1412.4553 [gr-qc]

85. Radice D, Perego A, Zappa F, Bernuzzi S (2018) GW170817: joint constraint on the neutron star equation of state from multimessenger observations. Astrophys J Lett 852:L29. https://doi.org/10.3847/2041-8213/aaa402. arXiv:1711.03647 [astro-ph.HE]
86. Radice D, Bernuzzi S, Del Pozzo W, Roberts LF, Ott CD (2017) Probing extreme-density matter with gravitational wave observations of binary neutron star merger remnants. Astrophys J Lett 842:L10. https://doi.org/10.3847/2041-8213/aa775f. arXiv:1612.06429 [astro-ph.HE]
87. Perego A, Bernuzzi S, Radice D (2019) Thermodynamics conditions of matter in neutron star mergers. Eur Phys J A 55:124. https://doi.org/10.1140/epja/i2019-12810-7. arXiv:1903.07898 [gr-qc]
88. Radice D, Rezzolla L (2012) THC: a new high-order finite-difference high-resolution shock-capturing code for special-relativistic hydrodynamics. Astron Astrophys 547:A26. https://doi.org/10.1051/0004-6361/201219735. arXiv:1206.6502 [astro-ph.IM]
89. Radice D, Rezzolla L, Galeazzi F (2014) High-order fully general-relativistic hydrodynamics: new approaches and tests. Class Quant Grav 31:075012. https://doi.org/10.1088/0264-9381/31/7/075012. arXiv:1312.5004 [gr-qc]
90. Radice D, Rezzolla L, Galeazzi F (2014) Beyond second-order convergence in simulations of binary neutron stars in full general-relativity. Mon Not Roy Astron Soc 437:L46–L50. https://doi.org/10.1093/mnrasl/slt137. arXiv:1306.6052 [gr-qc]
91. Schneider AS, Roberts LF, Ott CD (2017) Open-source nuclear equation of state framework based on the liquid-drop model with Skyrme interaction. Phys Rev C 96:065802. https://doi.org/10.1103/PhysRevC.96.065802. arXiv:1707.01527 [astro-ph.HE]
92. Tootle SD, Papenfort LJ, Most ER, Rezzolla L (2021) Quasi-universal Behavior of the Threshold Mass in Unequal-mass, Spinning Binary Neutron Star Mergers. Astrophys J Lett 922:L19. https://doi.org/10.3847/2041-8213/ac350d. arXiv:2109.00940 [gr-qc]
93. Köppel S, Bovard L, Rezzolla L (2019) A general-relativistic determination of the threshold mass to prompt collapse in binary neutron star mergers. Astrophys J Lett 872:L16. https://doi.org/10.3847/2041-8213/ab0210. arXiv:1901.09977 [gr-qc]
94. Zappa F, Bernuzzi S, Radice D, Perego A, Dietrich T (2018) Gravitational-wave luminosity of binary neutron stars mergers. Phys Rev Lett 120:111101. https://doi.org/10.1103/PhysRevLett.120.111101. arXiv:1712.04267 [gr-qc]
95. Lindblom L (2018) Causal representations of neutron-star equations of state. Phys Rev D 97:123019. https://doi.org/10.1103/PhysRevD.97.123019. arXiv:1804.04072 [astro-ph.HE]
96. Lindblom L (2010) Spectral representations of neutron-star equations of state. Phys Rev D 82:103011 arXiv:1009.0738 [astro-ph.HE]
97. Carney MF, Wade LE, Irwin BS (2018) Comparing two models for measuring the neutron star equation of state from gravitational-wave signals. Phys Rev D 98:063004. https://doi.org/10.1103/PhysRevD.98.063004. arXiv:1805.11217 [gr-qc]
98. Abbott BP et al (2018) (LIGO Scientific, Virgo), GW170817: Measurements of neutron star radii and equation of state. Phys Rev Lett 121:161101. https://doi.org/10.1103/PhysRevLett.121.161101. arXiv:1805.11581 [gr-qc]
99. Romani RW, Kandel D, Filippenko AV, Brink TG, Zheng W (2022) PSR J0952–0607: the fastest and heaviest known galactic neutron star. Astrophys J Lett 934:L17. https://doi.org/10.3847/2041-8213/ac8007. arXiv:2207.05124 [astro-ph.HE]
100. Manchester RN, Hobbs GB, Teoh A, Hobbs M (2005) The Australia Telescope National Facility pulsar catalogue. Astron J 129:1993. https://doi.org/10.1086/428488. arXiv:astro-ph/0412641
101. Hannam M, Schmidt P, Bohé A, Haegel L, Husa S, Ohme F, Pratten G, Pürrer M (2014) Simple model of complete processing black-hole-binary gravitational waveforms. Phys Rev Lett 113:151101. https://doi.org/10.1103/PhysRevLett.113.151101. arXiv:1308.3271 [gr-qc]
102. Dietrich T, Bernuzzi S, Tichy W (2017) Closed-form tidal approximants for binary neutron star gravitational waveforms constructed from high-resolution numerical relativity simulations. Phys Rev D 96:121501. https://doi.org/10.1103/PhysRevD.96.121501. arXiv:1706.02969 [gr-qc]

103. Dietrich T et al (2019) Matter imprints in waveform models for neutron star binaries: tidal and self-spin effects. Phys Rev D 99:024029. https://doi.org/10.1103/PhysRevD.99.024029. arXiv:1804.02235 [gr-qc]
104. Dietrich T, Samajdar A, Khan S, Johnson-McDaniel NK, Dudi R, Tichy W (2019) Improving the NRTidal model for binary neutron star systems. Phys Rev D 100:044003. https://doi.org/10.1103/PhysRevD.100.044003. arXiv:1905.06011 [gr-qc]
105. Poisson E (1998) Gravitational waves from inspiraling compact binaries: The quadrupole-moment term. Phys Rev D 57:5287–5290. https://doi.org/10.1103/PhysRevD.57.5287
106. Arun KG, Buonanno A, Faye G, Ochsner E (2009) Higher-order spin effects in the amplitude and phase of gravitational waveforms emitted by inspiraling compact binaries: Ready-to-use gravitational waveforms. Phys Rev D 79:104023. https://doi.org/10.1103/PhysRevD.79.104023. [Erratum: Phys.Rev.D 84, 049901 (2011)], arXiv:0810.5336 [gr-qc]
107. Mikoczi B, Vasuth M, Gergely LA (2005) Self-interaction spin effects in inspiralling compact binaries. Phys Rev D 71:124043. https://doi.org/10.1103/PhysRevD.71.124043. arXiv:astro-ph/0504538
108. Bohé A, Faye G, Marsat S, Porter EK (2015) Quadratic-in-spin effects in the orbital dynamics and gravitational-wave energy flux of compact binaries at the 3PN order. Class Quant Grav 32:195010. https://doi.org/10.1088/0264-9381/32/19/195010. arXiv:1501.01529 [gr-qc]
109. Mishra CK, Kela A, Arun KG, Faye G (2016) Ready-to-use post-Newtonian gravitational waveforms for binary black holes with nonprecessing spins: an update. Phys Rev D 93:084054. https://doi.org/10.1103/PhysRevD.93.084054. arXiv:1601.05588 [gr-qc]
110. Bohé A et al (2017) Improved effective-one-body model of spinning, nonprecessing binary black holes for the era of gravitational-wave astrophysics with advanced detectors. Phys Rev D 95:044028. https://doi.org/10.1103/PhysRevD.95.044028. arXiv:1611.03703 [gr-qc]
111. Husa S, Khan S, Hannam M, Pürrer M, Ohme F, Jiménez Forteza X, Bohé A (2016) Frequency-domain gravitational waves from nonprecessing black-hole binaries I New numerical waveforms and anatomy of the signal. Phys Rev D 93:044006. https://doi.org/10.1103/PhysRevD.93.044006. arXiv:1508.07250 [gr-qc]
112. Nagar A et al (2018) Time-domain effective-one-body gravitational waveforms for coalescing compact binaries with nonprecessing spins, tides and self-spin effects. Phys Rev D 98:104052. https://doi.org/10.1103/PhysRevD.98.104052. arXiv:1806.01772 [gr-qc]
113. Akcay S, Bernuzzi S, Messina F, Nagar A, Ortiz N, Rettegno P (2019) Effective-one-body multipolar waveform for tidally interacting binary neutron stars up to merger. Phys Rev D 99:044051 arXiv:1812.02744 [gr-qc]
114. Hinderer T et al (2019) Distinguishing the nature of comparable-mass neutron star binary systems with multimessenger observations: GW170817 case study. Phys Rev D 100:06321. https://doi.org/10.1103/PhysRevD.100.063021. arXiv:1808.03836 [astro-ph.HE]

Chapter 8
Conclusions

The application of gravitational wave (GW) detections to probe and understand the dynamics of compact binaries in the strong gravity regime is dependent on two ingredients (i) construction of a bank of waveform templates to analyse the data observed in the detectors through match filtering and (ii) performing a Bayesian analysis on the detectors' data using these waveform templates to infer the properties of the system. Consequently, this thesis is divided into two parts; (i) construction of an analytical waveform model within the EOB formalism for massless scalar-tensor theories, and (ii) improving inference of properties of neutron stars using the GR-based waveform approximants for our analysis.

The aim of the Part I of the thesis is to analytically model the GW sources in scalar tensor theories. The analytical waveform template banks used by LIGO are either based on the PN formalism for inspiral and NR for merger and ringdown, as for example in the `IMRPhenom` waveform models or the NR-informed EOB formalism-based waveform models called `TEOBResumS` or `SEOBNR` to describe the complete dynamics. The waveform templates based on the EOB formalism were *important for robust parameter estimation* for the detection of the first GW event GW150914. Therefore, to robustly describe the binary dynamics and waveforms in the strong field regime for modified theories of gravity its post-Newtonian results should be recast within the EOB description of the two-body problem. In Part I, we mapped the PN information of the two-body problem in massless scalar-tensor theories within the EOB formalism. To achieve this, we first mapped the conservative dynamics into the EOB formalism building upon the PN expanded Lagrangian of the two-body problem at 3PN order for both the local-in-time (instantaneous) and nonlocal-in-time (tail contributions) in Chaps. 2 and 3, respectively. Since the finite size effects (tidal effects) also start at 3PN order for scalar-tensor theories, we also computed the contributions to the EOB Hamiltonian due to tidal effects for *circular* orbits in Chap. 2. As the EOB Hamiltonian derived in these two chapters describes the conservative dynamics of *general* binary orbits up to 3 PN order, we used this information to derive the scattering angle for unbound orbits in massless scalar-tensor theories

T. Jain, *Surfing the Gravitational Waves*, Springer Theses,
https://doi.org/10.1007/978-3-031-96180-9_8

up to 3 PN order in Chap. 4. Then we extended the radiation reaction force within the EOB formalism for bound orbits up to 1 PN order for *general* orbits. However, as a first step, we derived the angular momentum flux up to 1.5 PN order within the PN formalism for the two-body problem in Chap. 5 in harmonic coordinates. We then used this information along with the energy flux from the literature for scalar-tensor theories to derive the radiation reaction force in Chap. 6 within EOB formalism for generic orbits up to 1 PN order by focusing only on the tensorial contribution.

The current status of our work is that we have implemented the theoretical results obtained in Chaps. 2 and 3 with quasi-circular information for radiation reaction and waveforms from Ref. [1] into the EOB based waveform model `TEOBResumS` and the preliminary results based on our model showing the impact of the scalar-tensor corrections with respect to GR is presented in Fig. 8.1. The first natural extension of our work would be to use the GW data for parameter estimation analysis using this scalar-tensor theory dependent waveform model and to also integrate our model in LIGO waveform pipelines for future GW detections. The next step would then be to decompose the results derived in Chap. 6 in the form of quasi-circular orbit radiation reaction terms, i.e. total radiation reaction as a product of quasi-circular orbit and eccentric orbit radiation reaction terms, and then incorporate these corrections in `TEOBResumS` to construct the generic orbit waveform model for scalar-tensor theory. As the black-hole-neutron-star like binaries could be an important source to infer the scalar-tensor parameters, therefore another interesting extension of our work could be to first theoretically study spin-orbit interactions in scalar-tensor theories at higher PN orders and then recasting this information within the EOB formalism to be used for waveform construction for spinning binaries.

Additionally, there has been recent progress in using the scattering amplitude approach to compute the dynamics of the two-body system in GR in the post-Minkowskian effective-field-theory (PMEFT) formalism [2, 3]. As each PM order encapsulates the information contained at all the PN orders, there is a hope that the information computed using EFT tools can be used in waveform construction for bound orbits to exploit the physics from future GW detections. Therefore, we anticipate that such an effort would be extended to scalar-tensor theories of gravity in future where the results of Chap. 4 could be used to verify the scattering angle results computed in the PM formalism within its PN approximation.

The GW observations are also used to infer the parameters of the source and this forms the basis of the Part II of this thesis where we consider the GR-based waveform approximants to infer the source properties. In Part II, we used the non-prompt collapse hypothesis for the GW signal GW170817 and the mass-ration dependent fits for the threshold mass to study the possible improvement in the inference of NS properties. We considered the LVC posterior samples for the GW170817 signal for our analysis using GR-based waveform approximants. From our study, we observed that the non-prompt collapse hypothesis based on the mass-ratio dependent fits removes a significant part of the EOS parameter space and hence could be used to improve the constraints on the NS properties for similar EM-bright events in the future. An interesting extension of this work could be to use these fits to derive the constraint on the maximum mass for non-rotating NSs and compare it with results

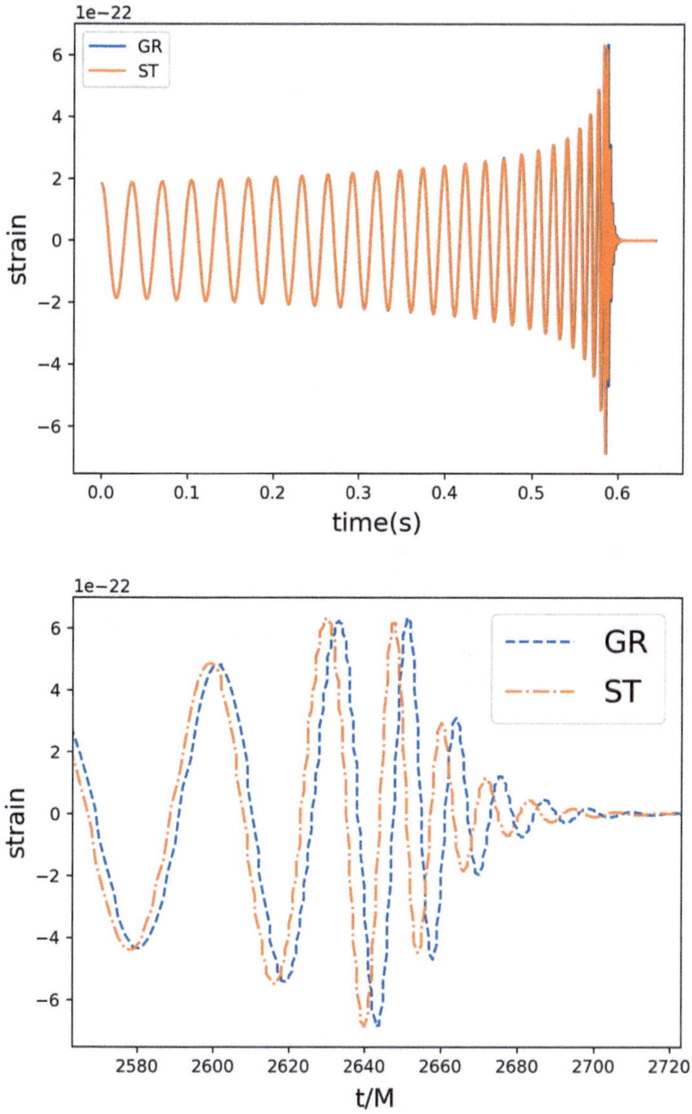

Fig. 8.1 Plot of the gravitational wave strain as a function of time in seconds for the full waveform in the upper panel and time in geometric units around the merger in the lower panel using `TEOBResumS` waveform model for scalar-tensor (ST) theory and general relativity (GR) for the parameters of GW150914 event, i.e. total mass $M = 45M_{\odot}$ and mass ratio $q = 1.25$. The blue and orange colours indicate ST and GR-based waveforms, respectively

obtained in Ref. [4]. The other possible investigation would be to use both the mass-ratio and spin-dependent threshold mass fits derived in Ref. [5] with the non-prompt collapse hypothesis on the GW170817 signal to study the impact of spins and to put constraints on the EOS.

References

1. Sennett N, Marsat S, Buonanno A (2016) Gravitational waveforms in scalar-tensor gravity at 2PN relative order. Phys Rev D 94:084003. arXiv:1607.01420. https://doi.org/10.1103/PhysRevD.94.084003 [gr-qc]
2. Kälin G, Porto RA (2020) Post-Minkowskian effective field theory for conservative binary dynamics. JHEP 11:106. arXiv:2006.01184. https://doi.org/10.1007/JHEP11(2020)106 [hep-th]
3. Mogull G, Plefka J, Steinhoff J (2020) Classical black hole scattering from a worldline quantum field theory. JHEP 02:048. arXiv:2010.02865. https://doi.org/10.1007/JHEP02(2021)048 [hep-th]
4. Kashyap R et al (2021) Numerical relativity simulations of prompt collapse mergers: threshold mass and phenomenological constraints on neutron star properties after GW170817. arXiv:2111.05183 [astro-ph.HE]
5. Tootle SD, Papenfort LJ, Most ER, Rezzolla L (2021) Quasi-universal behavior of the threshold mass in unequal-mass, spinning binary neutron star mergers. Astrophys J Lett 922:L19. arXiv:2109.00940. https://doi.org/10.3847/2041-8213/ac350d [gr-qc]

MIX
Papier aus verantwortungsvollen Quellen
Paper from responsible sources
FSC® C105338

If you have any concerns about our products,
you can contact us on
ProductSafety@springernature.com

In case Publisher is established outside the EU,
the EU authorized representative is:
Springer Nature Customer Service Center GmbH
Europaplatz 3, 69115 Heidelberg, Germany

Printed by Libri Plureos GmbH
in Hamburg, Germany